U·X·L
ENCYCLOPEDIA OF
BIOMES

U·X·L
ENCYCLOPEDIA OF
BIOMES

MARLENE WEIGEL

1

CONIFEROUS FOREST
CONTINENTAL MARGIN
DECIDUOUS FOREST
DESERT

WALLA WALLA COMMUNITY
COLLEGE LIBRARY

U·X·L®

AN IMPRINT OF THE GALE GROUP

DETROIT · SAN FRANCISCO · LONDON
BOSTON · WOODBRIDGE, CT

U•X•L Encyclopedia of Biomes
Marlene Weigel

STAFF

Julie L. Carnagie, *U•X•L Editor*
Carol DeKane Nagel, *U•X•L Managing Editor*
Thomas L. Romig, *U•X•L Publisher*
Meggin Condino, *Senior Analyst, New Product Development*

Margaret Chamberlain, *Permissions Specialist* (Pictures)

Rita Wimberley, *Senior Buyer*
Evi Seoud, *Assistant Production Manager*
Dorothy Maki, *Manufacturing Manager*

Michelle DiMercurio, *Art Director*
Cynthia Baldwin, *Product Design Manager*

Graphix Group, *Typesetting*

Library of Congress Cataloging-in-Publication Data

U•X•L encyclopedia of biomes / Marlene Weigel, Julie L. Carnagie, editor.
    p.   cm.
    Includes bibliographical references and indexes.
    Contents: v. 1. Coniferous forests, continental margins, deciduous forests, and deserts –
    v. 2. Grasslands, lakes and ponds, oceans, and rainforests – v. 3. Rivers, seashores, tundras, and wetlands.
    ISBN 0-7876-3732-7 (set). -- ISBN 0-7876-3733-5 (vol. 1). – ISBN 0-7876-3734-3
    (vol. 2). – ISBN 0-7876-3735-1 (vol. 3)
    1. Biotic communities Juvenile literature. [1. Biotic communities.] I. Weigel, Marlene. II.
    Carnagie, Julie. III. Title: Encyclopedia of biomes.
    QH541.14.U18 1999
    577.8'2–DC21
99-23395   CIP

This publication is a creative work fully protected by all applicable copyright laws, as well as by misappropriation, trade secret, unfair competition, and other applicable laws. The author and editors of this work have added value to the underlying factual material herein through one or more of the following: unique and original selection, coordination, expression, arrangement, and classification of the information. All rights to this publication will be vigorously defended.

Copyright © 2000

U•X•L, an Imprint of The Gale Group
27500 Drake Rd.
Farmington Hills, MI 48331-3535

All rights reserved, including the right of reproduction in whole or in part in any form.

Printed in the United States of America
10 9 8 7 6 5 4 3 2

TABLE OF
CONTENTS

VOLUME 3

READER'S GUIDE

U•X•L Encyclopedia of Biomes offers readers comprehensive and easy-to-use information on twelve of the Earth's major biomes and their many component ecosystems. Arranged alphabetically across three volumes, each biome chapter includes an overview; a description of how the biomes are formed; their climate; elevation; growing season; plants, animals, and endangered species; food webs; human culture; and economy. The information presented may be used in a variety of subject areas, such as biology, geography, anthropology, and current events. Each chapter includes a "spotlight" feature focusing on specific geographical areas related to the biome being discussed and concludes with a section composed of books, periodicals, Internet addresses, and environmental organizations for readers to conduct more extensive research.

ADDITIONAL FEATURES

Each volume of *U•X•L Encyclopedia of Biomes* includes a color-coded world biome map and a sixteen-page color insert. More than120 black-and-white photos, illustrations, and maps enliven the text, while sidebar boxes highlight fascinating facts and related information. All three volumes include a glossary; a bibliography; and a subject index covering all the subjects discussed in *U•X•L Encyclopedia of Biomes.*

NOTE

There are many different ways to describe certain aspects of a particular biome, and it would be impossible to include all of the classifications in *U•X•L Encyclopedia of Biomes.* However, in cases where more than one classification seemed useful, more than one was given. Please note that the classifications represented here may not be those preferred by all specialists in a particular field.

Every effort was made in this set to include the most accurate information pertaining to sizes and other measurements. Great variations exist in the available data, however. Sometimes differences can be accounted for in terms of what was measured: the reported area of a lake, for example, may vary depending upon the point at which measuring began. Other differences may result from natural changes that took place between the time of one measurement and another. Further, other data may be questionable because reliable information has been difficult to obtain. This is particularly true for remote areas in developing countries, where funds for scientific research are lacking and non-native scientists may not be welcomed.

ACKNOWLEDGMENTS

Special thanks are due for the invaluable comments and suggestions provided by the *U•X•L Encyclopedia of Biomes* advisors:

- Nancy Bard, Librarian, Thomas Jefferson High School for Science and Technology, Alexandria, Virginia

- Frances L. Cohen, Consultant, Malvern, Pennsylvania

- Valerie Doud, Science Teacher, Peru Junior High School, Peru, Indiana

- Elaine Ezell, Library Media Specialist, Bowling Green Junior High School, Bowling Green, Ohio

The author and editor would like to thank contributing writer Rita Travis for her work on the Coniferous Forest, Grassland, Tundra, and Wetland chapters.

COMMENTS AND SUGGESTIONS

We welcome your comments on this work as well as your suggestions for topics to be featured in future editions of *U•X•L Encyclopedia of Biomes*. Please write: Editors, *U•X•L Encyclopedia of Biomes*, U•X•L, 27500 Drake Rd., Farmington Hills, MI 48331-3535; call toll-free: 1-800-877-4253; fax: 248-699-8097; or send e-mail via www.galegroup.com.

WORDS TO KNOW

A

Abyssal plain: The flat midportion of the ocean floor that begins beyond the continental rise.

Acid rain: A mixture of water vapor and polluting compounds in the atmosphere that falls to the Earth as rain or snow.

Active margin: A continental margin constantly being changed by earthquake and volcanic action.

Aerial roots: Plant roots that dangle in midair and absorb nutrients from their surroundings rather than from the soil.

Algae: Plantlike organisms that usually live in watery environments and depend upon photosynthesis for food.

Algal blooms: Sudden increases in the growth of algae on the ocean's surface.

Alluvial fan: A fan-shaped area created when a river or stream flows downhill, depositing sediment into a broader base that spreads outward.

Amphibians: Animals that spend part, if not most, of their lives in water.

Amphibious: Able to live on land or in water.

Angiosperms: Trees that bear flowers and produce their seeds inside a fruit; deciduous and rain forest trees are usually angiosperms.

Annuals: Plants that live for only one year or one growing season.

Aquatic: Having to do with water.

Aquifier: Rock beneath the Earth's surface in which groundwater is stored.

Arachnids: Class of animals that includes spiders and scorpions.

Arctic tundra: Tundra located in the far north, close to or above the Arctic Circle.

Arid: Dry.

Arroyo: The dry bed of a stream that flows only after rain; also called a wash or a *wadi.*

Artifacts: Objects made by humans, including tools, weapons, jars, and clothing.

Artificial grassland: A grassland created by humans.

Artificial wetland: A wetland created by humans.

Atlantic blanket bogs: Blanket bogs in Ireland that are less than 656 feet (200 meters) above sea level.

Atolls: Ring-shaped reefs formed around a lagoon by tiny animals called corals.

B

Bactrian camel: The two-humped camel native to central Asia.

Bar: An underwater ridge of sand or gravel formed by tides or currents that extends across the mouth of a bay.

Barchan dunes: Sand dunes formed into crescent shapes with pointed ends created by wind blowing in the direction of their points.

Barrier island: An offshore island running parallel to a coastline that helps shelter the coast from the force of ocean waves.

Barrier reef: A type of reef that lines the edge of a continental shelf and separates it from deep ocean water. A barrier reef may enclose a lagoon and even small islands.

Bathypelagic zone: An oceanic zone based on depth that ranges from 3,300 to 13,000 feet (1,000 to 4,000 meters).

Bathyscaphe: A small, manned, submersible vehicle that accommodates several people and is able to withstand the extreme pressures of the deep ocean.

Bay: An area of the ocean partly enclosed by land; its opening into the ocean is called a mouth.

Beach: An almost level stretch of land along a shoreline.

Bed: The bottom of a river or stream channel.

Benthic: Term used to describe plants or animals that live attached to the seafloor.

Biodiverse: Term used to describe an environment that supports a wide variety of plants and animals.

Bio-indicators: Plants or animals whose health is used to indicate the general health of their environment.

Biological productivity: The growth rate of life forms in a certain period of time.

Biome: A distinct, natural community chiefly distinguished by its plant life and climate.

Blanket bogs: Shallow bogs that spread out like a blanket; they form in areas with relatively high levels of annual rainfall.

Bog: A type of wetland that has wet, spongy, acidic soil called peat.

Boreal forest: A type of coniferous forest found in areas bordering the Arctic tundra. Also called taiga.

Boundary layer: A thin layer of water along the floor of a river channel where friction has stopped the flow completely.

Brackish water: A mixture of freshwater and saltwater.

Braided stream: A stream consisting of a network of interconnecting channels broken by islands or ridges of sediment, primarily mud, sand, or gravel.

Branching network: A network of streams and smaller rivers that feeds a large river.

Breaker: A wave that collapses on a shoreline because the water at the bottom is slowed by friction as it travels along the ocean floor and the top outruns it.

Browsers: Herbivorous animals that eat from trees and shrubs.

Buoyancy: Ability to float.

Buran: Strong, northeasterly wind that blows over the Russian steppes.

Buttresses: Winglike thickenings of the lower trunk that give tall trees extra support.

C

Canopy: A roof over the forest created by the foliage of the tallest trees.

Canyon: A long, narrow valley between high cliffs that has been formed by the eroding force of a river.

Carbon cycle: Natural cycle in which trees remove excess carbon dioxide from the air and use it during photosynthesis. Carbon is then returned to the soil when trees die and decay.

Carnivore: A meat-eating plant or animal.

Carrion: Decaying flesh of dead animals.

Cay: An island formed from a coral reef.

Channel: The path along which a river or stream flows.

Chemosynthesis: A chemical process by which deep-sea bacteria use organic compounds to obtain food and oxygen.

Chernozim: A type of temperate grassland soil; also called black earth.

Chinook: A warm, dry wind that blows over the Rocky Mountains in North America.

Chitin: A hard chemical substance that forms the outer shell of certain invertebrates.

Chlorophyll: The green pigment in leaves used by plants to turn energy from the Sun into food.

Clear-cutting: The cutting down of every tree in a selected area.

Climax forest: A forest in which only one species of tree grows because it has taken over and only that species can survive there.

Climbers: Plants that have roots in the ground but use hooklike tendrils to climb on the trunks and limbs of trees in order to reach the canopy, where there is light.

Cloud forest: A type of rain forest that occurs at elevations over 10,500 feet (3,200 meters) and that is covered by clouds most of the time.

Commensalism: Relationship between organisms in which one reaps a benefit from the other without either harming or helping the other.

Commercial fishing: Fishing done to earn money.

Coniferous trees: Trees, such as pines, spruces, and firs, that produce seeds within a cone.

Consumers: Animals in the food web that eat either plants or other animals.

Continental shelf: A flat extension of a continent that tapers gently into the sea.

Continental slope: An extension of a continent beyond the continental shelf that dips steeply into the sea.

Convergent evolution: When distantly related animals in different parts of the world evolve similar characteristics.

Coral reef: A wall formed by the skeletons of tiny animals called corals.

Coriolis Effect: An effect on wind and current direction caused by the Earth's rotation.

Crustaceans: Invertebrate animals that have hard outer shells.

Current: The steady flow of water in a certain direction.

D

Dambo: Small marsh found in Africa.

Dark zone: The deepest part of the ocean, where no light reaches.

Deciduous: Term used to describe trees, such as oaks and elms, that lose their leaves during cold or very dry seasons.

Decompose: The breaking down of dead plants and animals in order to release nutrients back into the environment.

Decomposers: Organisms that feed on dead organic materials, releasing nutrients into the environment.

Dehydration: Excessive loss of water from the body.

Delta: Muddy sediments that have formed a triangular shape over the continental shelf near the mouth of a river.

Deposition: The carrying of sediments by a river from one place to another and depositing them.

Desalination: Removing the salt from seawater.

Desert: A very dry area receiving no more than 10 inches (25 centimeters) of rain during a year and supporting little plant or animal life.

Desertification: The changing of fertile lands into deserts through destruction of vegetation (plant life) or depletion of soil nutrients. Topsoil and groundwater are eventually lost as well.

Desert varnish: A dark sheen on rocks and sand believed to be caused by the chemical reaction between overnight dew and minerals in the soil.

Diatom: A type of phytoplankton with a geometric shape and a hard, glasslike shell.

Dinoflagellate: A type of phytoplankton having two whiplike attachments that whirl in the water.

Discharge: The amount of water that flows out of a river or stream into another river, a lake, or the ocean.

Doldrums: Very light winds near the equator that create little or no movement in the ocean.

Downstream: The direction toward which a river or stream is flowing.

Drainage basin: All the land area that supplies water to a river or stream.

Dromedary: The one-humped, or Arabian, camel.

Drought: A long, extremely dry period.

Dune: A hill or ridge of sand created by the wind.

Duricrusts: Hard, rocklike crusts on ridges that are formed by a chemical reaction caused by the combination of dew and minerals such as limestone.

E

Ecosystem: A network of organisms that have adapted to a particular environment.

Eddy: A current that moves against the regular current, usually in a circular motion.

Elevation: The height of an object in relation to sea level.

Emergent plants: Plants that are rooted at the bottom of a body of water that have top portions that appear to be above the water's surface.

Emergents: The very tallest trees in the rain forest, which tower above the canopy.

Engineered wood: Manufactured wood products composed of particles of several types of wood mixed with strong glues and preservatives.

Epilimnion: The layer of warm or cold water closest to the surface of a large lake.

Epipelagic zone: An oceanic zone based on depth that reaches down to 650 feet (200 meters).

Epiphytes: Plants that grow on other plants or hang on them for physical support.

Ergs: Arabian word for vast seas of sand dunes, especially those found in the Sahara.

Erosion: Wearing away of the land.

Estivation: An inactive period experienced by some animals during very hot months.

Estuary: The place where a river traveling through lowlands meets the ocean in a semi-enclosed area.

Euphotic zone: The zone in a lake where sunlight can reach.

Eutrophication: Loss of oxygen in a lake or pond because increased plant growth has blocked sunlight.

F

Fast ice: Ice formed on the surface of the ocean between pack ice and land.

Faults: Breaks in the Earth's crust caused by earthquake action.

Fell-fields: Bare rock-covered ground in the alpine tundra.

Fen: A bog that lies at or below sea level and is fed by mineral-rich ground-water.

First-generation stream: The type of stream on which a branching network is based; a stream with few tributaries. Two first-generation streams join to form a second-generation stream and so on.

Fish farms: Farms in which fish are raised for commercial use; also called hatcheries.

Fjords: Long, narrow, deep arms of the ocean that project inland.

Flash flood: A flood caused when a sudden rainstorm fills a dry riverbed to overflowing.

Floating aquatic plant: A plant that floats either partly or completely on top of the water.

Flood: An overflow caused when more water enters a river or stream than its channel can hold at one time.

Floodplain: Low-lying, flat land easily flooded because it is located next to streams and rivers.

Food chain: The transfer of energy from organism to organism when one organism eats another.

Food web: All of the possible feeding relationships that exist in a biome.

Forbs: A category of flowering, broadleaved plants other than grasses that lack woody stems.

Forest: A large number of trees covering not less than 25 percent of the area where the tops of the trees interlock, forming a canopy at maturity.

Fossil fuels: Fuels made from oil and gas that formed over time from sediments made of dead plants and animals.

Fossils: Remains of ancient plants or animals that have turned to stone.

Freshwater lake: A lake that contains relatively pure water and relatively little salt or soda.

Freshwater marsh: A wetland fed by freshwater and characterized by poorly drained soil and plant life dominated by nonwoody plants.

Freshwater swamp: A wetland fed by freshwater and characterized by poorly drained soil and plant life dominated by trees.

Friction: The resistance to motion when one object rubs against another.

Fringing reef: A type of coral reef that develops close to the land; no lagoon separates it from the shore.

Frond: A leaflike organ found on all species of kelp plants.

Fungi: Plantlike organisms that cannot make their own food by means of photosynthesis; instead they grow on decaying organic matter or live as parasites on a host.

G

Geyser: A spring heated by volcanic action. Some geysers produce enough steam to cause periodic eruptions of water.

Glacial moraine: A pile of rocks and sediments created as a glacier moves across an area.

Global warming: Warming of the Earth's climate that may be speeded up by air pollution.

Gorge: A deep, narrow pass between mountains.

Grassland: A biome in which the dominant vegetation is grasses rather than trees or tall shrubs.

Grazers: Herbivorous animals that eat low-growing plants such as grass.

Ground birds: Birds that hunt food and make nests on the ground or close to it.

Groundwater: Freshwater stored in rock layers beneath the ground.

Gulf: A large area of the ocean partly enclosed by land; its opening is called a strait.

Gymnosperms: Trees that produce seeds that are often collected together into cones; most conifers are gymnosperms.

Gyre: A circular or spiral motion.

H

Hadal zone: An oceanic zone based on depth that reaches from 20,000 to 35,630 feet (6,000 to 10,860 meters).

Hardwoods: Woods usually produced by deciduous trees, such as oaks and elms.

Hatcheries: Farms in which fish are raised for commercial use; also called fish farms.

Headland: An arm of land made from hard rock that juts out into the ocean after softer rock has been eroded away by the force of tides and waves.

Headwaters: The source of a river or stream.

Herbicides: Poisons used to control weeds or any other unwanted plants.

Herbivore: An animal that eats only plant matter.

Herders: People who raise herds of animals for food and other needs; they may also raise some crops but are usually not dependent upon them.

Hermaphroditic: Term used to describe an animal or plant in which reproductive organs of both sexes are present in one individual.

Hibernation: An inactive period experienced by some animals during very cold months.

High tide: A rising of the surface level of the ocean caused by the Earth's rotation and the gravitational pull of the Sun and Moon.

Holdfast: A rootlike structure by which kelp plants anchor themselves to rocks or the seafloor.

Hummocks: 1. Rounded hills or ridges, often heavily wooded, that are higher than the surrounding area; 2. Irregularly shaped ridges formed when large blocks of ice hit each other and one slides on top of the other; also called hammocks.

Humus: The nutrient-rich, spongy matter produced when the remains of plants and animals are broken down to form soil.

Hunter-gatherers: People who live by hunting animals and gathering nuts, berries, and fruits; normally, they do not raise crops or animals.

Hurricane: A violent tropical storm that begins over the ocean.

Hydric soil: Soil that contains a lot of water but little oxygen.

Hydrologic cycle: The manner in which molecules of water evaporate, condense as clouds, and return to the Earth as precipitation.

Hydrophytes: Plants that are adapted to grow in water or very wet soil.

Hypolimnion: The layer of warm or cold water closest to the bottom of a large lake.

Hypothermia: A lowering of the body temperature that can result in death.

I

Insecticides: Poisons that kill insects.

Insectivores: Plants and animals that feed on insects.

Intermittent stream: A stream that flows only during certain seasons.

Interrupted stream: A stream that flows aboveground in some places and belowground in others.

Intertidal zone: The seashore zone covered with water during high tide and dry during low tide; also called the middle, or the littoral, zone.

Invertebrates: Animals without a backbone.

K

Kelp: A type of brown algae that usually grows on rocks in temperate water.

Kettle: A large pit created by a glacier that fills with water and becomes a pond or lake.

Kopjes: Small hills made out of rocks that are found on African grasslands.

L

Labrador Current: An icy Arctic current that mixes with warmer waters off the coast of northeastern Canada.

Lagoon: A large pool of seawater cut off from the ocean by a bar or other landmass.

Lake: A usually large body of inland water that is deep enough to have two distinct layers based on temperature.

Latitude: A measurement on a map or globe of a location north or south of the equator. The measurements are made in degrees, with the equator, or dissecting line, being zero.

Layering: Tree reproduction that occurs when a branch close to the ground develops roots from which a new tree grows.

Levee: High bank of sediment deposited by a very silty river.

Lichens: Plantlike organisms that are combinations of algae and fungi. The algae produces the food for both by means of photosynthesis.

Limnetic zone: The deeper, central region of a lake or pond where no plants grow.

Littoral zone: The area along the shoreline that is exposed to the air during low tide; also called intertidal zone.

Longshore currents: Currents that move along a shoreline.

Lowland rain forest: Rain forest found at elevations up to 3,000 feet (900 meters).

Low tide: A lowering of the surface level of the ocean caused by the Earth's rotation and the gravitational pull of the Sun and Moon.

M

Macrophytic: Term used to describe a large plant.

Magma: Molten rock from beneath the Earth's crust.

Mangrove swamp: A coastal saltwater swamp found in tropical and subtropical areas.

Marine: Having to do with the oceans.

Marsh: A wetland characterized by poorly drained soil and by plant life dominated by nonwoody plants.

Mature stream: A stream with a moderately wide channel and sloping banks.

Meandering stream: A stream that winds snakelike through flat countryside.

Mesopelagic zone: An oceanic zone based on depth that ranges from 650 to 3,300 feet (200 to 1,000 meters).

Mesophytes: Plants that live in soil that is moist but not saturated.

Mesophytic: Term used to describe a forest that grows where only a moderate amount of water is available.

Mid-ocean ridge: A long chain of mountains that lies under the World Ocean.

Migratory: Term used to describe animals that move regularly from one place to another in search of food or to breed.

Mixed-grass prairie: North American grassland with a variety of grass species of medium height.

Montane rain forest: Mountain rain forest found at elevations between 3,000 and 10,500 feet (900 and 3,200 meters).

Mountain blanket bogs: Blanket bogs in Ireland that are more than 656 feet (200 meters) above sea level.

Mouth: The point at which a river or stream empties into another river, a lake, or an ocean.

Muck: A type of gluelike bog soil formed when fully decomposed plants and animals mix with wet sediments.

Muskeg: A type of wetland containing thick layers of decaying plant matter.

Mycorrhiza: A type of fungi that surrounds the roots of conifers, helping them absorb nutrients from the soil.

N

Neap tides: High tides that are lower and low tides that are higher than normal when the Earth, Sun, and Moon form a right angle.

Nekton: Animals that can move through the water without the help of currents or wave action.

Neritic zone: That portion of the ocean that lies over the continental shelves.

Nomads: People or animals who have no permanent home but travel within a well-defined territory determined by the season or food supply.

North Atlantic Drift: A warm ocean current off the coast of northern Scandinavia.

Nutrient cycle: Natural cycle in which mineral nutrients are absorbed from the soil by tree roots and returned to the soil when the tree dies and the roots decay.

O

Oasis: A fertile area in the desert having a water supply that enables trees and other plants to grow there.

Ocean: The large body (or bodies) of saltwater that covers more than 70 percent of the Earth's surface.

Oceanography: The exploration and scientific study of the oceans.

Old stream: A stream with a very wide channel and banks that are nearly flat.

Omnivore: Organism that eats both plants and animals.

Ooze: Sediment formed from the dead tissues and waste products of marine plants and animals.

Oxbow lake: A curved lake formed when a river abandons one of its bends.

Oxygen cycle: Natural cycle in which the oxygen taken from the air by plants and animals is returned to the air by plants during photosynthesis.

P

Pack ice: A mass of large pieces of floating ice that have come together on an open ocean.

Pamir: A high altitude grassland in Central Asia.

Pampa: A tropical grassland found in South America.

Pampero: A strong, cold wind that blows down from the Andes Mountains and across the South American pampa.

Pantanal: A wet savanna that runs along the Upper Paraguay River in Brazil.

Parasite: An organism that depends upon another organism for its food or other needs.

Passive margin: A continental margin free of earthquake or volcanic action, and in which few changes take place.

Peat: A type of soil formed from slightly decomposed plants and animals.

Peatland: Wetlands characterized by a type of soil called peat.

Pelagic zone: The water column of the ocean.

Perennials: Plants that live at least two years or seasons, often appearing to die but returning to "life" when conditions improve.

Permafrost: Permanently frozen topsoil found in northern regions.

Permanent stream: A stream that flows continually, even during a long, dry season.

Pesticides: Poisons used to kill anything that is unwanted and is considered a pest.

Phosphate: An organic compound used in making fertilizers, chemicals, and other commercial products.

Photosynthesis: The process by which plants use the energy from sunlight to change water (from the soil) and carbon dioxide (from the air) into the sugars and starches they use for food.

Phytoplankton: Tiny, one-celled algae that float on ocean currents.

Pingos: Small hills formed when groundwater freezes.

Pioneer trees: The first trees to appear during primary succession; they include birch, pine, poplar, and aspen.

Plankton: Plants or animals that float freely in ocean water; from the Greek word meaning "wanderer."

Plunge pool: Pool created where a waterfall has gouged out a deep basin.

Pocosin: An upland swamp whose only source of water is rain.

Polar climate: A climate with an average temperature of not more than 50°F (10°C) in July.

Polar easterlies: Winds occurring near the Earth's poles that blow in an easterly direction.

Pollination: The carrying of pollen from the male reproductive part of a plant to the female reproductive part of a plant so that reproduction may occur.

Polygons: Cracks formed when the ground freezes and contracts.

Pond: A body of inland water that is usually small and shallow and has a uniform temperature throughout.

Pool: A deep, still section in a river or stream.

Prairie: A North American grassland that is defined by the stature of grass groups contained within (tall, medium, and short).

Prairie potholes: Small marshes no more than a few feet deep.

Precipitation: Rain, sleet, or snow.

Predatory birds: Carnivores that hunt food by soaring high in the sky to obtain a "bird's-eye" view and then swoop down to capture their prey.

Primary succession: Period of plant growth that begins when nothing covers the land except bare sand or soil.

Producers: Plants and other organisms in the food web of a biome that are able to make food from nonliving materials, such as the energy from sunlight.

Profundal zone: The zone in a lake where no more than 1 percent of sunlight can penetrate.

Puna: A high-altitude grassland in the Andes Mountains of South America.

R

Rain forest: A tropical forest, or jungle, with a warm, wet climate that supports year-round tree growth.

Raised bogs: Bogs that grow upward and are higher than the surrounding area.

Rapids: Fast-moving water created when softer rock has been eroded to create many short drops in the channel; also called white water.

Reef: A ridge or wall of rock or coral lying close to the surface of the ocean just off shore.

Rhizomes: Plant stems that spread out underground and grow into a new plant that breaks above the surface of the soil or water.

Rice paddy: A flooded field in which rice is grown.

Riffle: A stretch of rapid, shallow, or choppy water usually caused by an obstruction, such as a large rock.

Rill: Tiny gully caused by flowing water.

Riparian marsh: Marsh usually found along rivers and streams.

Rip currents: Strong, dangerous currents caused when normal currents moving toward shore are deflected away from it through a narrow channel; also called riptides.

River: A natural flow of running water that follows a well-defined, permanent path, usually within a valley.

River system: A river and all its tributaries.

S

Salinity: The measure of salts in ocean water.

Salt lake: A lake that contains more than 0.1 ounce of salt per quart (3 grams per liter) of water.

Salt pan: The crust of salt left behind when a salt or soda lake or pond dries up.

Saltwater marsh: A wetland fed by saltwater and characterized by poorly drained soil and plant life dominated by nonwoody plants.

Saltwater swamp: A wetland fed by saltwater and characterized by poorly drained soil and plant life dominated by trees.

Saturated: Soaked with water.

Savanna: A grassland found in tropical or subtropical areas, having scattered trees and seasonal rains.

Scavenger: An animal that eats decaying matter.

School: Large gathering of fish.

Sea: A body of saltwater smaller and shallower than an ocean but connected to it by means of a channel; sea is often used interchangeably with ocean.

Seafloor: The ocean basins; the area covered by ocean water.

Sea level: The height of the surface of the sea. It is used as a standard in measuring the heights and depths of other locations such as mountains and oceans.

Seamounts: Isolated volcanoes on the ocean floor that do not break the surface of the ocean.

Seashore: The strip of land along the edge of an ocean.

Secondary succession: Period of plant growth occurring after the land has been stripped of trees.

Sediments: Small, solid particles of rock, minerals, or decaying matter carried by wind or water.

Seiche: A wave that forms during an earthquake or when a persistent wind pushes the water toward the downwind end of a lake.

Seif dunes: Sand dunes that form ridges lying parallel to the wind; also called longitudinal dunes.

Shelf reef: A type of coral reef that forms on a continental shelf having a hard, rocky bottom. A shallow body of water called a lagoon may be located between the reef and the shore.

Shoals: Areas where enough sediments have accumulated in the river channel that the water is very shallow and dangerous for navigation.

Shortgrass prairie: North American grassland on which short grasses grow.

Smokers: Jets of hot water expelled from clefts in volcanic rock in the deep-seafloor.

Soda lake: A lake that contains more than 0.1 ounce of soda per quart (3 grams per liter) of water.

Softwoods: Woods usually produced by coniferous trees.

Sonar: The use of sound waves to detect objects.

Soredia: Algae cells with a few strands of fungus around them.

Source: The origin of a stream or river.

Spit: A long narrow point of deposited sand, mud, or gravel that extends into the water.

Spores: Single plant cells that have the ability to grow into a new organism.

Sport fishing: Fishing done for recreation.

Spring tides: High tides that are higher and low tides that are lower than normal because the Earth, Sun, and Moon are in line with one another.

Stagnant: Term used to describe water that is unmoving and contains little oxygen.

Star-shaped dunes: Dunes created when the wind comes from many directions; also called stellar dunes.

Steppe: A temperate grassland found mostly in southeast Europe and Asia.

Stone circles: Piles of rocks moved into a circular pattern by the expansion of freezing water.

Straight stream: A stream that flows in a straight line.

Strait: The shallow, narrow channel that connects a smaller body of water to an ocean.

Stream: A natural flow of running water that follows a temporary path that is not necessarily within a valley; also called brook or creek. Scientists often use the term to mean any natural flow of water, including rivers.

Subalpine forest: Mountain forest that begins below the snow line.

Subduction zone: Area where pressure forces the seafloor down and under the continental margin, often causing the formation of a deep ocean trench.

Sublittoral zone: The seashore's lower zone, which is underwater at all times, even during low tide.

Submergent plant: A plant that grows entirely beneath the water.

Subpolar gyre: The system of currents resulting from winds occurring near the poles of the Earth.

Subsistence fishing: Fishing done to obtain food for a family or a community.

Subtropical: Term used to describe areas bordering the equator in which the weather is usually warm.

Subtropical gyre: The system of currents resulting from winds occurring in subtropical areas.

Succession: The process by which one type of plant or tree is gradually replaced by others.

Succulents: Plants that appear thick and fleshy because of stored water.

Sunlit zone: The uppermost part of the ocean that is exposed to light; it reaches down to about 650 feet (200 meters) deep.

Supralittoral zone: The seashore's upper zone, which is never underwater, although it may be frequently sprayed by breaking waves; also called the splash zone.

Swamp: A wetland characterized by poorly drained soil, stagnant water, and plant life dominated by trees.

Sward: Fine grasses that cover the soil.

Swell: Surface waves that have traveled for long distances and become more regular in appearance and direction.

T

Taiga: Coniferous forest found in areas bordering the Arctic tundra; also called boreal forest.

Tallgrass prairie: North American grassland on which only tall grass species grow.

Tannins: Chemical substances found in the bark, roots, seeds, and leaves of many plants and used to soften leather.

Tectonic action: Movement of the Earth's crust, as during an earthquake.

Temperate bog: Peatland found in temperate climates.

Temperate climate: Climate in which summers are hot and winters are cold, but temperatures are seldom extreme.

Temperate zone: Areas in which summers are hot and winters are cold but temperatures are seldom extreme.

Thermal pollution: Pollution created when heated water is dumped into the ocean. As a result, animals and plants that require cool water are killed.

Thermocline: Area of the ocean's water column, beginning at about 1,000 feet (300 meters), in which the temperature changes very slowly.

Thermokarst: Shallow lakes in the Arctic tundra formed by melting permafrost; also called thaw lakes.

Tidal bore: A surge of ocean water caused when ridges of sand direct the ocean's flow into a narrow river channel, sometimes as a single wave.

Tidepools: Pools of water that form on a rocky shoreline during high tide and that remain after the tide has receded.

Tides: Rhythmic movements, caused by the Earth's rotation and the gravitational pull of the Sun and Moon, that raise or lower the surface level of the oceans.

Tombolo: A bar of sand that has formed between the beach and an island, linking them together.

Trade winds: Winds occurring both north and south of the equator to about 30 degrees latitude; they blow primarily east.

Transverse dunes: Sand dunes lying at right angles to the direction of the wind.

Tree: A large woody perennial plant with a single stem, or trunk, and many branches.

Tree line: The elevation above which trees cannot grow.

Tributary: A river or stream that flows into another river or stream.

Tropical: Term used to describe areas close to the equator in which the weather is always warm.

Tropical tree bog: Bog found in tropical climates in which peat is formed from decaying trees.

Tropic of Cancer: A line of latitude about 25 degrees north of the equator.

Tropic of Capricorn: A line of latitude about 25 degrees south of the equator.

Tsunami: A huge wave or upwelling of water caused by undersea earthquakes that grows to great heights as it approaches shore.

Tundra: A cold, dry, windy region where trees cannot grow.

Turbidity current: A strong downward-moving current along the continental margin caused by earthquakes or the settling of sediments.

Turbine: An energy-producing engine.

Tussocks: Small clumps of vegetation found in marshy tundra areas.

Typhoon: A violent tropical storm that begins over the ocean.

U

Understory: A layer of shorter, shade-tolerant trees that grow under the forest canopy.

Upstream: The direction from which a river or stream is flowing.

Upwelling: The rising of water or molten rock from one level to another.

V

Veld: Temperate grassland in South Africa.

Venom: Poison produced by animals such as certain snakes and spiders.

Vertebrates: Animals with a backbone.

W

Wadi: The dry bed of a stream that flows only after rain; also called a wash or an *arroyo*.

Warm-bodied fish: Fish that can maintain a certain body temperature by means of a special circulatory system.

Wash: The dry bed of stream that flows only after rain; also called an *arroyo* or a *wadi*.

Water column: All of the waters of the ocean, exclusive of the sea bed or other landforms.

Water cycle: Natural cycle in which trees help prevent water runoff, absorb water through their roots, and release moisture into the atmosphere through their leaves.

Waterfall: A cascade of water created when a river or stream falls over a cliff or erodes its channel to such an extent that a steep drop occurs.

Water table: The level of groundwater.

Waves: Rhythmic rising and falling movements in the water.

Westerlies: Winds occurring between 30 degrees and 60 degrees latitude; they blow in a westerly direction.

Wet/dry cycle: A period during which wetland soil is wet or flooded followed by a period during which the soil is dry.

Wetlands: Areas that are covered or soaked by ground or surface water often enough and long enough to support plants adapted for life under those conditions.

Wet meadows: Freshwater marshes that frequently dry up.

X

Xerophytes: Plants adapted to life in dry habitats or in areas like salt marshes or bogs.

Y

Young stream: A stream close to its headwaters that has a narrow channel with steep banks.

Z

Zooplankton: Animals, such as jellyfish, corals, and sea anemones, that float freely in ocean water.

CONIFEROUS FOREST

A tree is a large woody plant with one main stem, or trunk, and many branches that lives year after year. A forest is a large number of trees covering not less than 25 percent of an area where the tops of the trees, called crowns, interlock forming an enclosure or canopy at maturity. This chapter is about coniferous (koh-NIH-fuhr-uhs) forests. Almost all coniferous trees, such as pines and firs bear their seeds inside cones. Most also have stiff, flattened or needle-like leaves that usually remain green all winter. For information about other kinds of forests, see the chapters titled "Deciduous Forest" and "Rain Forest."

More than 50 percent of the world's coniferous forests are found in Asia, primarily in Siberia, China, Korea, and Japan, and on the slopes of the Himalaya and Hindu Kush Mountains. In Europe they cover much of Scandinavia and the coast of the Baltic Sea and are found on the primary mountain ranges—the Alps, the Vosges, and the Carpathians. In North America, coniferous forests stretch across the northern part of the continent from Alaska to Newfoundland; down into Washington, Oregon, and California; along the Cascades, the Sierra Nevadas, and the Rocky Mountains; eastward through portions of the north central states, through New Jersey and into Maine. In the southern states, they are found in portions of Virginia, North Carolina, South Carolina, Georgia, Alabama, Mississippi, Louisiana, Arkansas, Florida, and Tennessee. In the Southern Hemisphere, they are found in Mexico, along the western coast of South America, in parts of Argentina and Brazil, in parts of the Australasia region, and in portions of Africa.

HOW CONIFEROUS FORESTS DEVELOP

The first forests evolved during Earth's prehistoric past. Since then, all forests have developed in essentially the same way, by means of a process called succession.

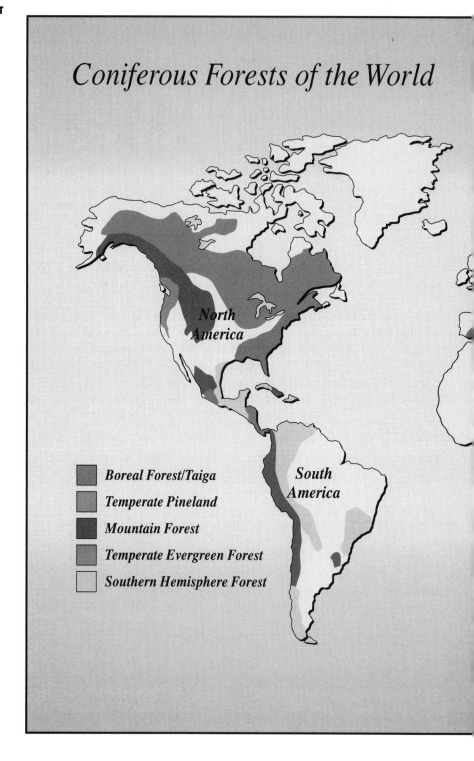

Coniferous Forests of the World

North America

South America

Boreal Forest/Taiga

Temperate Pineland

Mountain Forest

Temperate Evergreen Forest

Southern Hemisphere Forest

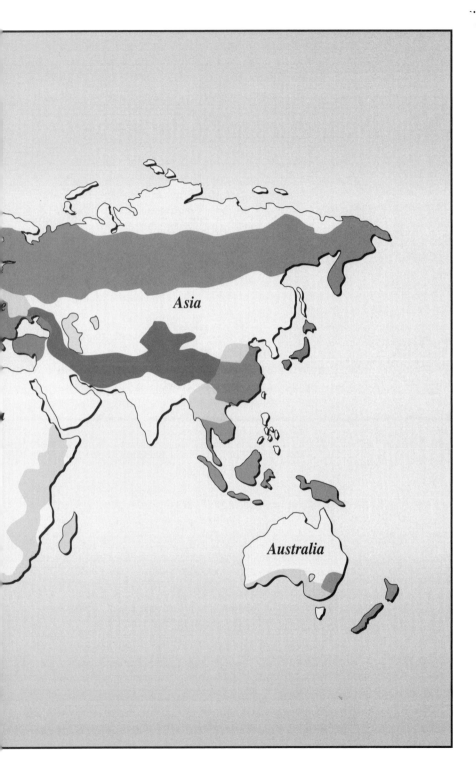

THE FIRST FORESTS

The first forests evolved from ferns and other prehistoric plants, which, over time having adapted to the surrounding environment, grew more tree-like. Trees that preferred a warm, humid, tropical climate developed first, followed by those that gradually adapted to drier, cooler weather. Coniferous forests flourished during the Jurassic Period, about 190,000,000 years ago. Fossils (ancient remains that have turned to stone) of the Walachia tree, an ancestor of the conifers that grew over 250,000,000 years ago, show that it resembled trees found growing in the Southern Hemisphere today.

About 1,000,000 years ago, during the great Ice Ages, glaciers (slow-moving masses of ice) covered much of the planet and destroyed many of the world's forests. By the time the glaciers finally retreated, about 10,000 years ago, they had scoured much of the land of plants and soil, leaving only bare rock. Birch trees were among the first trees to return to the regions once covered by ice. In fact, the years between 8000 and 3000 B.C. are often referred to as the "Birch Period." Birches prepared the way for other trees.

SUCCESSION

Trees compete with one another for sunlight, water, warmth and nutrients, and a forest is constantly changing. The process by which one type of plant or tree is gradually replaced by others is called succession. Succession took place following the last Ice Age, just as it takes place today when the land is stripped of vegetation from other causes, such as forest fires. Succession produces different types of forests in different regions, but the process is essentially the same everywhere. During succession, different species of trees become dominant.

Primary succession Primary succession usually begins on bare soil or sand where no plants grew before. When the right amount of sunlight, moisture, and air temperature are present, seeds begin to germinate (grow). These first plants are usually made up of grasses and forbs (a nonwoody broad-leaved plant) type. They continue to grow and eventually form meadows. Over time and as conditions change, other plants, such as shrubs and trees, begin to grow. These plants become dominant and replace the grasses and forbs.

A JOB IN THE WOODS

Forestry is the profession which deals with development and management of forests. The main objective is to ensure that there will always be trees and a supply of timber, but foresters are also involved with the conservation of soil, water, and wildlife resources, and with preserving land for recreation. By the mid-1990s, there were more than 20,000 foresters in the United States, and about 50 colleges and universities offered degrees in the field.

Silviculture is a branch of forestry dealing with forest growth. Foresters in this branch cultivate different types of forests and encourage them to grow as quickly as possible using such methods as fertilizers. Some forests are developed for lumber, paper, or pulp. Researchers use scientific breeding methods to raise the type of tree best for each purpose.

With a need for superior quality trees grown for specific uses, cloning may become a tool of the forester in the future. A clone is the genetically identical copy of an organism, and an entire forest could be cloned from one ideal parent tree.

Stages in Forest Succession

Lichens

FIRST STAGE:
Bare rocks with
lichens

Ferns

Mosses

Grasses

SECOND STAGE:
Ferns, mosses
and grasses

THIRD STAGE:
Pioneer trees
dominate

FOURTH STAGE:
Deciduous trees
become dominant

CLIMAX STAGE:
All one Species

*An illustration showing
the five stages in
forest succession, from
the bare rock stage to
the climax forest
stage.*

As primary succession continues, "pioneer" trees—birch, pine, poplar, and aspen—begin to thrive. They are all tall, sun-loving trees, and they quickly take over the meadow. However, they also change the environment by making shade. Now, trees with broad leaves, such as red oaks, that prefer some protection from the Sun, begin to take root. If conditions are right, a mixed forest of sun-loving and shade-loving trees may continue for many years. Eventually, however, more changes occur.

The climax forest Because seedlings from pioneer trees do not grow well in shade, pioneer trees are not replaced as the older trees die from old age, disease, and other causes. This results in broad-leaved trees becoming dominant. However, the shade from some broad-leaved trees can be so dense even their own seedlings cannot grow. As a result, seedlings from trees that prefer heavy shade, such as beech and sugar maple, begin to thrive and eventually take over the forest. These trees produce such deep shade that only trees that can live in deep shade will succeed there. The result is a climax forest in which only one species of tree becomes dominant. If one tree dies, another of the same species grows to take its place. In this way, a climax forest can continue for thousands of years. However, few true climax forests actually exist

---

FIRE IN THE FOREST

About 12,000,000 acres (4,800,000 hectares) of forest in the United States are damaged or destroyed by fire every year. Forest fires can spread as fast as 10 miles (15 kilometers) per hour downwind. Most forest fires are caused by humans and occur in coniferous forests, where both the air and the wood tend to be dry. Also, pine needles contain resin (sap), which burns easily.

Not only do fires burn trees, they also destroy other plant and animal life. Ground fires burn up organic material beneath the vegetation on the forest floor. Although ground fires move slowly, they do much damage. Fires that burn small vegetation and loose material are called surface fires. A crown fire is one that moves quickly through the tops of trees or shrubs.

Fires can actually benefit the forest, however. The huge fire in Yellowstone National Park in 1988 is an example. In the beginning, it was a controlled fire. In other words, the park's managers allowed it to burn naturally so that the natural balance of the forest could be restored. Although the fire eventually got out of control, it only destroyed about 20 percent of the park. In its wake, it left ashes rich in minerals such as calcium and phosphorous. These minerals promoted the growth of new plants, such as pine grass, which grows beneath Douglas firs and flowers only after a fire. The standing dead trees now attract certain types of birds, like the woodpecker and tree swallow. Also, the smoke and heat killed harmful insects and parasitic fungi that endangered trees. The comparatively few animals that died provided food for other species. In recent years, many of the burned areas have begun to support tree seedlings, as well as many other plants.

---

because other changes take place that interfere with the forest's stability. Fires, floods, high winds, and people can destroy acres of trees, while glaciers can mow them down, and volcanoes can smother them with ash and lava or knock them over with explosive force. Then the trees must start over.

Secondary succession When the land has been stripped of trees, it will eventually be covered with them again if the area is left alone. This is called secondary succession. Because some soil usually remains intact, secondary succession takes place more quickly than primary succession. Seeds from other forests in neighboring regions are blown by the wind or carried by animals to the site. Soon, the seeds take root, seedlings sprout, and the process begins again.

> ### A CONIFER FOR ST. NICK
> The most popular tree used as a Christmas tree is the coniferous fir. Fir trees are cone-shaped and can grow between 30 and 150 feet (9 and 46 meters) tall. Their needles are very fragrant and remain on the tree for a long time after it has been cut. Most of the trees used for Christmas trees are not wild but are grown on tree plantations.

KINDS OF CONIFEROUS FORESTS

Forests can be classified in many ways. In general, however, coniferous forests are categorized as boreal/taiga, mountain, temperate evergreen, temperate pine, or Southern Hemisphere forests.

BOREAL FOREST/TAIGA

The word boreal (BOHR-ee-yuhl) means "northern," and taiga (TAY-guh) is a Russian word for "little sticks." Boreal forests, or taiga, are found in regions bordering the Arctic tundra (a region so dry and cold that no trees will grow). These are the great northern forests of Canada, Alaska, Russia, and Scandinavia, and they form some of the largest forest biomes in the world.

Trees in the boreal forest grow at higher latitudes (a distance north or south of the equator measured in degrees) than trees in any other type of forest. The most common are spruce, pine, and fir.

The boreal forest may be divided into three zones. The northernmost zone is the forest-tundra, where trees meet treeless land. Few species are able to survive here. The second zone is the lichen (LY-ken; an algae and fungi combination) woodland, or sparse taiga. Here, trees grow far apart and a lichen mat covers the forest floor. The southernmost region is the closed-canopy forest where many species of conifers grow close together, and the forest floor is covered with mosses that grow well in shade.

MOUNTAIN CONIFEROUS FOREST

Mountain coniferous forests are found on mountains below the permanent snowfields. They are located in the Rockies, Cascades, and Sierra

Nevadas of North America; the Alps and Carpathians in Europe; and the Hindu Kush and Himalayas in Asia.

Forests on the upper slopes just below the snowfields are called sub-alpine forests. At these higher elevations, where the weather is harshest, trees such as the bristlecone pine are long-lived but often stunted in growth. Lodgepole pines grow erect, but alpine firs grow close to the ground, as if shrinking from the harsh climate. Forests on the lower and middle slopes are called montane forests. Here, in the more moderate climate, conifers grow taller.

Trees common in mountain forests vary from region to region. In North America, Douglas fir, sierra redwood, and ponderosa pine predominate. Silver fir and larch are found in Europe, and chir pines, Himalayan firs, and morinda spruces grow in Asian forests.

TEMPERATE EVERGREEN FOREST

Temperate evergreen forests grow in regions with moderate, humid climates. Summers are warm and winters may be cool, but temperatures are

---

THE CEDAR—SYMBOL OF POWER AND LONG-LIFE

Cedar, one of the most durable and fragrant woods, has been used by humans since at least 2000 B.C. It does not rot easily and is resistant to fire and high winds. Great cedar forests were once found in Algeria, Morocco, the Himalayas, the Pacific Northwest in the United States, and in Lebanon in the Middle East. The Phoenicians, who lived in Lebanon, were excellent carpenters and sold much of their wood.

Some of the Phoenicians' best customers were the ancient Egyptians who lived in the Nile Valley where few trees grow. Because cedar is so strong, the Egyptians used it to construct homes and other buildings and for wooden rollers that moved huge stones into place for the construction of the pyramids. Cedar sawdust was used in the process of mummifying the dead, and the resin was used for embalming and coating coffins.

Cedar was also used in the construction of King Solomon's temple in Jerusalem between 965 to 926 B.C. The Romans used it to build ships, and the Roman emperor Hadrian brought about legislation to conserve the cedars of Lebanon between A.D. 117 and 138.

The North American western red cedar was used by coastal Native Americans. Clothing and baskets were made from bark, rope from the branches, and canoes and lodges from the trunks. Cedars have also long been valued in Japan, where conservation was first practiced during the sixteenth century. Now these trees form stands more than 1,000 years old.

Although most of the once mighty cedar forests are gone, today cedar is still used for building boats and homes. Its most common household use is in clothes closets and chests, where the aroma of cedar is said to repel cloth-eating moths.

seldom extreme. In North America, these forests range along the Pacific coast from Alaska and British Columbia to west central California, where they are sometimes called temperate rain forests. Common tree species include western hemlock, western red cedar, Douglas fir, and coast redwood.

The giant redwoods of California and southern Oregon are the tallest trees in America, many exceeding 300 feet (91 meters) in height and measuring up to 40 feet (12 meters) in circumference. The tallest individual tree measures 368 feet (112 meters) and has a circumference of 44 feet (13 meters). A species of redwoods found on the western slopes of the Sierra Nevada Mountains, the giant Sequoias, are the world's largest trees by volume and are also among the world's oldest trees. One Sequoia named "General Sherman" stands 272 feet (83 meters) tall, has a diameter of 37 feet (11 meters), and may be as much as 3,800 years old.

> **HOLD THE CALORIES!**
>
> Not to be outdone by the redwoods with their reputation as the world's tallest trees, the Mexican cypress has its own world record. An outstanding individual cypress in Tule, Mexico, known as "El Gigante," is the world's fattest tree. Although El Gigante is only 140 feet (43 meters) tall, its waistline measures 115 feet (35 meters) in circumference.

Where winters are mildest, temperate evergreen forests often contain broad-leaved evergreen trees. The humid conditions aid the growth of moss and other moisture-loving plants on the forest floor.

TEMPERATE PINE FOREST

Temperate pinelands grow in hilly or upland regions with warm, dry climates. In the United States, they are found in southern California and in a region stretching from New Jersey to southern Florida and west to Texas. They also occur in Mexico, China, and throughout the regions bordering the Mediterranean Sea.

SOUTHERN HEMISPHERE CONIFEROUS FOREST

Coniferous forests located primarily in the Southern Hemisphere, including those of Mexico, Central and South America, Africa, Southeast Asia, Australia, and New Zealand, are not the same as forests in the Northern Hemisphere. They may be classified as mountain, temperate evergreen, or temperate pineland forests, but the tree species are quite different. There are no native pines, firs, or spruces, although various northern species of pines may have been introduced for commercial purposes. In general, the native coniferous trees are smaller and their leaves have different forms. Most are found in the mountains or high plateaus.

For example, in South America, native conifers include the Parana pine, monkey-puzzle, Patagonian cedar, and Chilean cedar. In Australia, bunya pine, white cypress pine, and hoop pine grow. Forests of kauri pine grow in New Zealand, and the South African yellowwood is found in southern Africa. The Norfolk Island pine, native to the island of that name, is a popular ornamental tree worldwide.

CLIMATE

The climate of a coniferous forest depends upon where it is located. In general, the more northerly the latitude, the cooler the climate. The presence of mountain ranges and oceans also effects the climate of an area. In Japan, for example, Siberian air masses bring severe winters to some forests, while other forests are influenced by warm ocean currents and have a more tropical climate.

BOREAL FOREST/TAIGA

The most severe climate is found in the boreal forest, or taiga, where temperatures are below freezing for more than half of the year. Winter temperatures range from -65° to 30°F (-54° to -1°C), and summer temperatures from 20° to 70°F (-7° to 21°C). However, because the taiga is a land of extremes, temperatures can drop as low as -76°F (-60°C) in winter or climb as high as 104°F (40°C) in summer.

The greater portion of precipitation (rain, snow, or sleet) in the boreal forest comes from summer rain, which averages 12 to 33 inches (30 to 84 centimeters) per year.

MOUNTAIN CONIFEROUS FOREST

Mountain forests face cold, dry climates and high winds. The higher the elevation, the harsher the conditions. Scientists estimate that for every 300 feet (91 meters) in elevation, the temperature drops more than 1 degree. On Alaskan mountains, temperatures in January average about 8°F (-13.2°C) and in July only 47°F (8.2°C).

In general, northern hemisphere forests found on the northern side of mountains are shaded from the Sun, and the air is cooler. The forests receive more rainfall and have denser stands (groups) of trees and other plants. Forests on the southern side of mountains are drier and warmer and have less vegetation.

TEMPERATE EVERGREEN FOREST

The redwood and Pacific Northwestern forests have a climate that is moderated by the Pacific Ocean and the coastal mountain ranges. In the Olympic Rain Forest in Washington, for example, the temperature is always above freezing in winter and is seldom higher than 85°F (29°C) in summer. Up to 145 inches (368 centimeters) of rain fall here annually.

TEMPERATE PINE FOREST

In the Mediterranean and parts of California, winters are warm and wet and summers are hot and dry. Droughts (extremely dry periods) may be common. In the Mediterranean region, for example, winter temperatures usually do not fall below freezing.

SOUTHERN HEMISPHERE CONIFEROUS FOREST

The climate in Southern Hemisphere forests varies, depending upon where they are found. In the tropics (the regions around the equator), where the forests are located at higher elevations, clouds of mist may blanket them creating cool and damp conditions. In more temperate regions, such as in the mountains of Chile, conditions are drier and colder.

GEOGRAPHY OF CONIFEROUS FORESTS

The geography of coniferous forests includes landforms, elevation, soil, mineral resources, and water resources.

LANDFORMS

Coniferous forest landforms vary, depending upon the location, and may include mountains, valleys, rolling hills, or flat plateaus. The boreal forest landscape is dotted with wetlands, lakes, and ponds. Unique to the Canadian and Alaskan forest is the muskeg, a type of bog or marsh with thick layers of decaying plant matter. The muskeg looks like moss-covered ground but is so wet and spongy that hikers sink and trees often wobble in the unstable soil. When the muskeg freezes in winter, irregularly shaped ridges, called hummocks, may form.

Treeless hollows are often found in coniferous forests in the Northern Hemisphere. These hollows were usually created when glaciers moved over the area and destroyed the trees. The glaciers also left behind basins that eventually filled with water and became lakes. Streams and rivers are common in these forests, as well.

ELEVATION

Coniferous forests grow at different elevations throughout the world, from sea level (the average height of the surface of the sea) to more than 15,000 feet (4,572 meters) above sea level. Temperate rain forests along the northern Pacific coast are usually found below 2,700 feet (823 meters). The mountain forests in the Pacific Northwest range between 3,000 and 7,000 feet (914 and 2,134 meters), and those in the Rocky Mountains between 4,000 and 7,500 feet (1,219 and 2,286 meters). The giant Sequoias in the Sierra Nevada Mountains grow between 4,500 and 8,000 feet (1,372 and

2,439 meters). Subalpine forests in the Sierra Nevadas are found between 6,500 and 11,000 feet (1,981 and 3,353 meters) and in the Rockies between 7,500 and 11,500 feet (2,286 and 3,505 meters).

In Peru, mountain forests begin around 9,800 feet (2,987 meters) and subalpine forests grow to an altitude of 15,000 feet (4,572 meters). In central Japan, mountain forests are common above 6,560 feet (2,000 meters), however, on the island of Hokkaido, coniferous forests are found at sea level. Himalayan conifers dominate below 10,000 feet (3,048 meters) and Himalayan firs above that point.

SOIL

In general, the presence of trees protect soils from erosion by holding it in place with their roots. Fallen trees are important in conserving and cycling nutrients back into the soil and in reducing erosion. Trees also create windbreaks, which helps prevent topsoil from being blown away. Soil in a coniferous forest is acidic, because the needles of coniferous trees have a high acid content. These soils are also affected by both climate and location.

In the boreal forest, for example, soils are poorly developed because of the cold temperatures. A layer of permafrost (permanently frozen soil) prevents rain and melting snow from being absorbed deep into the ground. Because the moisture stays close to the surface, the topsoil is soggy. The cold is also responsible for the slow rate of decomposition of dead plants and animals, which means that soils form slowly, and there are not a lot of soil-mixing creatures.

Mountain soils are usually dry because the sloping terrain allows rain and melting snow to run off. Therefore, these soils often make up only a thin layer over a rocky foundation, and tree roots do not penetrate deeply. This runoff also washes away nutrients.

In temperate evergreen forests in North America, the soils are reddish in color and high in aluminum and iron. Decomposition takes place rapidly here because of the high moisture content. This richer soil supports more plant growth on the forest floor.

MINERAL RESOURCES

Mineral resources found deep in the ground below coniferous forests began developing millions of years ago. As ancient trees died and soil gradually built up over them, their remains were compressed and turned into coal, oil, and natural gas.

In northern Russian forests, coal, oil, and gas are found beneath the forest floor. In other parts of the taiga, aluminum is mined. In the province of Cita in far eastern Russia, mining and primary-ore processing dominate the economy. Gold, tin, tungsten, molybdenum, and lead are among the many minerals found here.

Gold was often mined from the rivers and streams that run through the forests of the Pacific northwest in the United States and Canada, and nickel is found in the northern forest of Manitoba, Canada.

WATER RESOURCES

In temperate regions, water resources include rivers, streams, springs, lakes, and ponds. Because permafrost prevents water from sinking very deeply into the soil, wetlands, lakes, and ponds form in the boreal forests. Mountain forests are often dotted with lakes and ponds that were carved out by glaciers. Mountains that were formed by volcanic action often have craters that filled with water and became lakes. Crater Lake in Oregon is an example of such a lake.

PLANT LIFE

Most forests contain a mixture of trees, and coniferous forests are no different. Stands of deciduous (dee-SID-joo-uhs) trees, such as larches, birches, and poplars, may exist within their boundaries. (Deciduous trees lose their leaves at the end of each growing season.) For more complete information on these kinds of trees, see the chapters titled "Deciduous Forest" and "Rain Forest."

These mixed trees and smaller plants grow to different heights, forming "layers" in the forest. The tallest trees create a canopy, or roof, over the rest. In the coniferous forest, these trees are often spruces and firs. Beneath the canopy grows the understory, a layer of shorter, shade-tolerant trees, such as the Pacific dogwoods of California. The next layer, only a few feet off the ground, is composed of small shrubs, such as junipers, blueberry, witch-hobble, and mountain laurel. Growing close to the ground are wild flowers, grasses, ferns, mosses, and mushrooms. In a coniferous forest, the canopy is so dense that little light is able to penetrate down to the forest floor. As a result of this limited light and high humidity, organisms such as algae, fungi, and lichens tend to flourish.

ALGAE, FUNGI, AND LICHENS

It is generally recognized that algae (AL-jee), fungi (FUHN-jee), and lichens do not fit neatly into the plant category. In this chapter, however, they will be discussed as if they, too, were plants.

Algae Most algae are single-celled organisms, although a few are multicellular. Many species have the ability to make their own food by means of photosynthesis (foh-toh-SIHN-thuh-sihs), the process by which plants use the energy from sunlight to change water and carbon dioxide into the sugars and starches they use for food. Others absorb nutrients from their surroundings.

Algae may reproduce in one of three ways. Some split into two or more parts, each part becoming a new, separate plant. Others form spores (single cells that have the ability to grow into a new organism). A few reproduce sexually, during which cells from two different plants unite to create a new plant.

Fungi Unlike algae, fungi cannot make their own food by means of photosynthesis. Some fungi, like mold and mushrooms, obtain nutrients from dead or decaying organic (derived from living organisms) matter. They assist in the decomposition (breaking down) of this matter and in releasing into the soil the nutrients needed by other plants. Other fungi are parasites and attach themselves to other living things. Fungi reproduce by means of spores.

One type of fungi, called mycorrhiza, surrounds the roots of conifers, helping them absorb nutrients from the soil. Slippery jack, a toadstool fungi that grows in coniferous forests, gets its name from the slimy material on its cap.

Lichens Lichens are combinations of algae and fungi that live in cooperation. The fungi surround the algal cells, and the algae produce food for themselves and the fungi by means of photosynthesis. It is believed that the fungi may help keep the algae moist. In harsher climates, such as that of the forest-tundra, lichens are often the only vegetation to survive. Because they have no root system, they can grow on bare rock, and, while their growth is slow, lichens often live for hundreds of years. Hemlock forests may contain as many as 150 species of lichens. Reindeer moss, a common type of lichen, is found in many coniferous forests.

Like algae, lichens can reproduce in several ways. If a spore from the fungus lands next to an alga, they can join together to form a new lichen. Lichens can also reproduce by means of soredia. Soredia are algal cells with a few strands of fungus around them. When soredia break off, they form new lichens wherever they land.

GREEN PLANTS OTHER THAN TREES

Most green plants need several basic things to grow: light, air, water, warmth, and nutrients. In the coniferous woodland, light, water and warmth are not always abundant. Nutrients, such as nitrogen, phosphorus, and potassium, are obtained from the soil, which may not always have a large supply. For this reason, plants beneath many coniferous trees are often sparse and must often make special adaptations in order to survive. The pitcher plant, for example, catches insects for food, while others fight with trees for space and nutrients by emitting chemicals into the soil that prevent the germination of trees.

Coniferous woodlands are home to both annual and perennial plants. Annuals live only one year or one growing season. Perennials live at least two years or two growing seasons, often appearing to die when the climate becomes too cold or dry, but returning to "life" when conditions improve.

Growing season Location, soil, and moisture all help to determine the number and type of green plants that grow in the forest. Many different plants are found in the temperate evergreen forests, where the rich soil, warm climate, and moisture promote growth year-round. Fewer plants are found in the colder, drier, boreal forests where the growing season may last only about 12 weeks. In the hot, dry climate of the Mediterranean forests, plant growth prospers during the winter rainy season.

Reproduction Most green plants, such as wildflowers, reproduce by means of pollination (the process in which pollen is carried by visiting animals or the wind from the male reproductive flower part to the female reproductive part. As the growing season comes to an end, seeds are usually produced. The seed's hard outer covering protects it during cold winters or long dry seasons until it sprouts.

Lichens growing on the bark of a Scots pine. (Reproduced by permission of Corbis. Photograph by Chris Mattison.)

A few woodland plants, such as ferns, reproduce by means of spores. Perennial grasses produce rhizomes, long, rootlike stems that spread out below ground. These stems develop their own root systems and send up sprouts that grow into new plants.

Common coniferous forest green plants other than trees Typical green plants found in coniferous forests include mosses, forbs, and ferns.

MOSSES Mosses make up the green carpet on the forest floor. They also grow on tree trunks, decaying logs, and rocks. Mosses range in size from microscopic varieties to those more than 40 inches (1 meter) long. Each moss plant is formed like a tiny tree, with a single straight trunk and leaves growing from it. Mosses grow close together and can store large quantities of water.

Plume moss, also called feather moss or boreal forest moss, resembles an ostrich plume. It forms dense, light-green mats on rocks, rotten wood, or peaty (highly rich, organic) soil, especially in mountain forests of the Northern Hemisphere.

Some mosses are epiphytes, plants that grow on other plants for physical support. Epiphytes are found in warmer climates, such as that of the Pacific Northwest. They absorb water and nutrients from rain and debris that collects on the supporting plants rather than from their own roots.

FORBS Forbs are flowering, broad-leaved, perennial plants. They have root systems that may reach as deep as 20 feet (6 meters), and, because they are able to survive long dry periods, some live as long as 50 years. Wildflowers are the most common forb.

Because little sunlight may reach the forest floor, some forbs store food in bulbs and rootstocks beneath the ground. Their flowers are usually small, grow low to the ground, and bloom before taller plants block all sunlight. Common forbs include trillium, bloodroot, spring beauty, Dutchman's-breeches, shooting stars, yellow violets, anemones, lilies, and delphiniums.

FERNS Many species of ferns grow in coniferous forests. Ferns are non-flowering perennial plants which reproduce by means of spores. Sword ferns are long and wispy, with toothed leaflets. Licorice ferns grow on tree trunks and stumps and are often seen draped over branches. Deer fern prefer the moist forest floor.

CONIFEROUS TREES

A tree is a woody perennial plant often living from 100 to 250 years. Some of the oldest living things on Earth are the bristlecone pines of the western United States, which may be as old as 5,000 years.

A bristlecone pine in Mount Evans, Colorado. These species of trees are the oldest in the world. (Reproduced by permission of Corbis. Photograph by Galen Rowell.)

Most coniferous trees have a single strong stem, or trunk. This gives them an advantage over smaller woody plants in that most of their growth is directed upward. Some, such as the redwoods, can reach nearly 400 feet (122 meters) in height, and their trunks may weigh over 1,000 tons (907 metric tons). While broad-leaved trees spread their limbs and branches out from the trunk to create a crown of leaves, conifers devote their energy to growing ever taller.

Each year, as a tree grows, its trunk is thickened with a new layer, or ring, of cells. When a tree is cut down, its age can be determined by how many of these rings are present, although those in conifers are not as pronounced. As a tree ages, the cells from the center outward become hardened to produce a sturdy core. Like all green plants, trees grow by means of photosynthesis, during which they release the oxygen essential to animals and humans back into the air.

In general, trees are divided into two groups according to how they bear their seeds. Angiosperms have flowers and produce their seeds inside a fruit, and many shed their leaves during cold or very dry periods. Broad-leaved trees, such as oaks and elms, are often angiosperms. Gymnosperms often produce seeds that grow together in cones. Most conifers are gymnosperms.

Conifers are well adapted to cool or cold temperatures and long dry periods. Their stiff, narrow needles are designed to prevent loss of moisture and offer less resistance to strong winter winds, which results in less wind damage. As a rule, coniferous trees do not shed their needles in autumn. They remain green all year and are often called "evergreens." They remain ready for the sudden warm weather and can take advantage of the short growing season. However, evergreens do shed the inner growth of needles approximately every two to three years, but because these needles were formed during past growth, their loss does not hinder new growth of the tree.

Growing season In general, forests need at least 10 to 15 inches (25 to 38 centimeters) of annual precipitation and at least 14 weeks of weather warm enough to promote growth. However, these conditions vary depending upon the location of the forest.

> ## THE OLDEST OF ALL LIVING THINGS?
>
> High in the Rocky Mountains of southwest California, Nevada, Utah, Arizona, New Mexico, and Colorado live the bristlecone pines, the oldest trees in the world. A tree cut down on Wheeler Ridge contained 4,900 rings of growth, and the oldest living specimen, the Methuselah Tree of California, is estimated at 4,600 years old. These twisted, ancient trees exist only at high altitudes—between 8,000 and 10,000 feet (2,432 and 3,040 meters)—and seldom grow more than 30 feet (9 meters) tall.

The growing season in the taiga, for example, is short, lasting only about 12 weeks. With little precipitation and cold temperatures, trees do not develop well. Here, even the oldest of trees is short and stunted. Sub-alpine forests also experience harsh climates, but their growing season is longer, enabling trees to grow taller.

Forests in more temperate climates enjoy more moisture, warmer weather, and a longer growing season. Because growing conditions are so favorable in the Pacific Northwest, trees like the redwoods grow to gigantic proportions. Warm pineland climates also produce rapid growth, as long as the rainy season is dependable.

Reproduction Coniferous plants bear female and male cones, many of which grow on the same tree. Male cones, which are small and soft, produce pollen, which fertilizes the eggs in the female cone. These eggs become seeds, which eventually fall from the cones and produce new trees.

Conifers also reproduce by layering. Layering occurs when a branch that is low to the ground is covered by soil. Roots form from the buried portion of the branch and grow into a new tree.

Common coniferous trees Common coniferous trees include parana pine, bunya pine, lodgepole pine, white pine, yew, Norway spruce, and yellowwood.

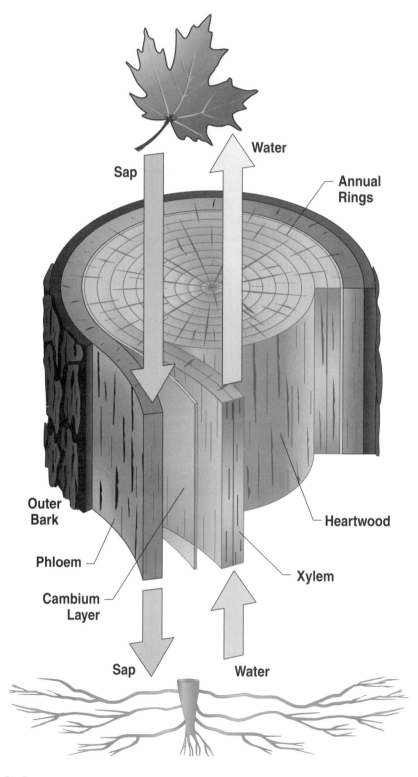

Water

Sap

Annual
Rings

Outer
Bark

Phloem

Cambium
Layer

Heartwood

Xylem

Sap

Water

*An illustration showing
the parts of a tree,
including the annual
rings and the direction
of water and sap.*

[18]

PARANA PINE In South America, the Parana pine, or candelabra tree, is an important conifer found in the hilly country of Brazil and Argentina. It is a slow-growing tree with a very straight trunk that reaches as high as 150 feet (46 meters). It is an important source of timber and, as a result, is being overharvested. Because it grows so slowly, the parana pine is being replaced by other types of faster-growing pines.

BUNYA PINE The bunya pine is a large Australian conifer that grows in the humid areas of Queensland. Reaching 100 feet (30 meters) or more in height, it is known for its large leafy crown and symmetrical branches. The wood is used for boxes, plywood, and veneers, and bunya seeds were once a food staple for the Australian Aborigines.

LODGEPOLE PINE The lodgepole pine is common in British Columbia, the Pacific Northwest, the Rocky Mountains, and along the Alaskan coast. Its needles are about 2 inches (5 centimeters) long and very sharp. Cones are yellowish brown and often grow in clusters of six or more. Because the tree is so straight, Native Americans used the trunk as the main support in their lodges, giving the tree its name.

WHITE PINE Because its trunk is large and the grain of the wood is soft and even, the white pine is a valuable North American timber tree. It often reaches almost 200 feet (61 meters) in height and has a trunk more than 3 feet (0.9 meter) in diameter. Eastern white pines once grew in large stands in Maine, Minnesota, and Manitoba, Canada. Western white pines once flourished in British Columbia and the northwestern United States; however, because of overharvesting, very few now remain.

YEW Yews grow between 10 and 60 feet (3 and 18 meters) tall, have reddish purple bark, and needles up to 1 inch (2.5 centimeters) long. They are found throughout North America and are part of the understory in the Pacific coast forests. The wood is strong and often used to build smaller, decorative items.

NORWAY SPRUCE Norway spruces have thick crowns and grow symmetrically, sometimes to heights of 200 feet (61 meters). Cones are long and tapered, and male cones produce so much pollen that the forest floor is often yellow with it. Norway spruces are common in Europe and Asia.

BROAD-LEAVED EVERGREENS—THE TREES-IN-BETWEEN

In northern climates, most broad-leaved trees lose their leaves in the autumn. However, some that live in climates where winters are mild but wet and summers are hot and dry remain green all year long. They are not conifers, because they do not produce seeds in cones, but they have learned to conserve moisture as conifers do by producing small leaves that they retain for several years. In effect, they are somewhere in between conifers and deciduous trees, having characteristics of both. Common species include the olive tree of the Mediterranean region, the eucalyptus of Australia and South America, the cork oak of Spain and Portugal, and the canyon live oak of California.

Certain broad-leaved evergreens have other unusual characteristics. For example, the cork oak produces bark as thick as 1 foot (30 centimeters), which can be harvested and used for corks. Another example is the eucalyptus, the giant tree of the southern hemisphere. Called the "giant gum" in Australia, some individuals may attain heights of 300 feet (91 meters), rivaling the giant Sequoias of the Northern Hemisphere.

YELLOWWOOD Yellowwoods once formed extensive forests in South Africa from Cape Province to Zululand and the Transvaal, but now exist only in patches. They grow tall for southern hemisphere conifers, reaching as high as 180 feet (55 meters), and their branches support many vines and creepers. These trees have narrow, yellow-green leaves, and their bark is grey and flaky.

ENDANGERED SPECIES

Trees can be threatened by natural dangers, such as forest fires, animals, and diseases, as well as by humans. Fires are more of a threat in dry climates, while animals and diseases seem prevalent in all climates. For example, when deer populations get too large, they can destroy forests by eating wildflowers and tree seedlings. Also, if enough insects attack a stand of trees, all the leaves are eaten and the trees die. Pollution is a serious threat because it appears to weaken trees, allowing pests and diseases to overtake them more easily.

In New Zealand, the kauri pine, once threatened by overharvesting, is now protected. However, forests in the Himalayan Mountains in Asia are currently being overharvested. In the European Alps, people have interfered with forest growth to such an extent that the trees now grow 500 feet (152 meters) lower on the mountain slopes than they did 1,000 years ago.

ANIMAL LIFE

Coniferous forests support a wide range of animals, and different regions are home to different species. These animals can be classified as microorganisms, invertebrates, amphibians, reptiles, birds, and mammals.

MICROORGANISMS

A microorganism is an animal, such as a protozoa, that cannot be seen without the aid of a microscope. Every forest is host to millions of these tiny creatures. Microscopic roundworms, or nematodes, for example, live by the thousands in small areas of soil in coniferous forests and aid the process of decomposition.

Bacteria Bacteria are microorganisms that are always present in woodland soil, where they help decompose dead plant and animal matter. In temperate climates, bacteria help create nutrient-rich humus (broken down organic matter). Fewer bacteria are at work in dry climates or in moist climates with long dry seasons.

Redwood trees in the Redwood National Park in California. In this park, logging of the redwoods is prohibited and fallen trees are left in order to protect the natural ecosystem. (Reproduced by Corbis. Photograph by Jim Sugar Photography.)

INVERTEBRATES

Animals without backbones are called invertebrates. They include simple animals such as worms, and more complex animals such as the wasp and the snail. Certain groups of invertebrates must spend part of their lives in water. Generally speaking, these types are not found in the trees, but in ponds, lakes, and streams, or in pools of rainwater.

Insects are found in abundance in coniferous forests, and they multiply very quickly when the weather warms up. Mosquitoes and midges swarm in clouds in boreal forests.

Food The green plants in the forest provide food for some insects, such as caterpillars and moths. Other insects, like wood ants and pine-bark beetles, eat dying or dead trees. Certain insects have very specific food requirements. Caterpillars of the Pandora moth feed on healthy pine trees, often killing them. White pine butterfly larvae feed only on the needles of the Douglas fir. Some insect larvae, however, store fat in their bodies and do not even have to look for food.

Snails prefer to eat plants, while bees, butterflies, and moths gather pollen and nectar (sweet liquid) from flowers. Arachnids (spiders), which are carnivores, prey on insects. If they are big enough, certain spiders will even eat small lizards, mice, and birds.

Reproduction The first part of an invertebrate's life cycle is spent as an egg. The second stage is the larva (such as a caterpillar), which may actually be divided into several stages between which the outer casing is shed as the animal increases in size. During the third, or pupal, stage, the animal's casing offers as much protection as an egg. An example of this stage is seen when a caterpillar spins a cocoon to live in while it develops into a butterfly. Finally, the adult emerges, usually by chewing its way out of its casing.

In coniferous forests, insect eggs and pupae are found on twigs, under the snow, or in cracks in trees. Caterpillars often emerge from eggs laid in the bark of trees. Some insects are picky about the trees they use. The white pine butterfly, for example, deposits its eggs only in Douglas firs. Pools of water on the damp forest floor provide a good location for the breeding of insects that require water, especially mosquitoes and midges.

Common coniferous forest invertebrates The pine panthea moth caterpillar and white pine wee-

A Parana pine, or candelabra tree, in South America. Because its an important source of timber, it is being overharvested. (Reproduced by permission of Corbis. Photograph by Carl Purcell.)

vil live on white pine trees. The caterpillar eats the needles and the weevil's larva eats the new shoots at the tips of stems. Because the weevil kills the shoots at the top, new shoots grow out of the sides of the stem instead, and the tree grows crookedly.

Engelmann spruce bark beetle, which bores under tree bark, lives in the high spruce forests in Colorado. At one time the beetle population became so numerous that millions of spruce trees were killed.

The caterpillar of the pine hawk moth does serious damage to European pine trees. However, the adult moth, which lives only a few days to mate and lay eggs, eats very little.

WE'RE BACK!

Lodgepole pines and jack pines are among the first trees to grow after a fire. Their seeds are sealed in cones by resins (sap) and may remain inactive for a long time on the forest floor. Some lodgepole pine seeds, for example, have been known to survive in this way for more than 80 years. Only the heat from a fire can melt the resin, thus releasing the seeds. The seeds quickly germinate in the ash-enriched soil, and a new forest begins to grow.

AMPHIBIANS

Amphibians are vertebrates (animals with backbones) that usually spend part, if not most, of their lives in water. Most amphibians are found in warm, moist, freshwater environments and in temperate zones (areas in which temperatures are seldom extreme). Amphibians found in coniferous forests include salamanders, newts, toads, and frogs. They generally live in moist areas like the underside of a log or beneath a mass of leaves. Only a few amphibians are found in the colder boreal forests.

Because amphibians breathe through their skin, and only moist skin can absorb oxygen, they must usually remain close to a water source, although mature animals leave the pools for dry land where they feed on both plants and insects. In warm climates, amphibians must find shade during the day or risk dying in the heat of the Sun.

Amphibians are cold-blooded animals, which means their bodies assume about the same temperature as their environment. Because they need a warm environment in order to be active, as temperatures get cooler, they slow down and seek shelter. During the winter season in temperate climates, they hibernate (remain inactive). In hot, dry climates, amphibians go through estivation, another inactive period similar to hibernation. While the soil is still moist from the rain, they dig themselves a foot or more into the ground, where they remain until the rains return. Only their nostrils remain exposed to the surface.

Food Amphibians use their long tongues to capture their prey. Even though some have teeth, they do not chew but swallow their food whole. Their larvae are mostly herbivorous, feeding on vegetation. Adult frogs and toads feed not only on algae and other plants, but also on insects, such as mosquitoes.

Reproduction Mating and egg-laying for most amphibians must take place in water. Male sperm are deposited in the water and swim to and penetrate the jelly-like eggs laid by the female. As the young develop into larvae and young adults, they often have gills for breathing that require a watery habitat. Once they mature, they develop lungs and can live on land.

Some amphibian females carry their eggs inside their bodies until they hatch. Certain species protect the eggs until the young are born, while others lay the eggs and abandon them.

Most amphibians reach maturity at three or four years. They breed for the first time about one year after they become adults.

Common coniferous forest amphibians The black salamander is commonly found in the high mountain forests in Europe, especially where Norway spruces grow. The black coat of the salamander absorbs heat and helps keep its body temperature up. The salamander is viviparous, (vy-VIP-ah-ruhs), meaning that the young are kept inside the mother's body until they are fully developed.

Many of this forests trees and plants will be destroyed by this forest fire. (Reproduced by permission of Field Mark Publications. Photograph by Robert J. Huffman.)

The Pacific tree frog, which lives in the Pacific coast coniferous forests, is small and slender with long legs and suction cups on the bottom of its toes that aid in climbing.

THE WEB OF LIFE: PREDATOR AND PREY

Some people shudder at the thought of predators killing deer. After all, the deer are so cute and those mountain lions and wolves so bloodthirsty! But a balance is needed between predators and prey or other problems arise. For example, in 1905, the Kaibab Forest in Arizona supported about 4,000 deer. Because the forest had the potential of supporting up to 30,000, people who wanted to enlarge the deer herd destroyed the predators that fed on the deer. Altogether they killed 7,000 coyotes, 716 mountain lions, and a number of wolves.

By 1918, however, the deer herds had increased to more than 40,000, and the plants the deer used for food began to suffer. By 1923, the herd had increased to 100,000, and many food plants had been completely wiped out. During the winter of 1924–25, 60,000 deer, many of which were fawns, died of starvation. The only solution was to restore the balance and return the predators to the forest.

REPTILES

Reptiles are cold-blooded vertebrates, such as lizards, turtles, and snakes, that depend on their environment for warmth. Therefore, they are usually most active when the weather is warm. No reptiles do well in extreme temperatures, either hot or cold. During hot, dry periods, they find some shade or a hole in which to wait for cooler weather. During chilly nights, they become slow and sluggish. In temperate climates, snakes may hibernate in burrows during the long winter.

Food Most reptiles are carnivorous. Snakes consume their prey whole—and often alive—without chewing. Their teeth often curve backward, which keeps their prey from escaping. It can take an hour or more to swallow a large victim.

The diet of lizards varies, depending upon the species. Some have long tongues with sticky tips and specialize in insects. Many are carnivores that eat small mammals and birds. The water they need is usually obtained from the food they eat. Turtles, for example, feed on soft plant material or small animals, or both.

Reproduction Most reptiles reproduce sexually, and their young come from eggs. The eggs are leathery and tough, and the offspring are seldom coddled. Some females remain with the eggs, but most reptiles bury the eggs in a hole and abandon them. The young are left to hatch by themselves. Once free of the eggs, the babies dig themselves out of the hole and begin life on their own. Many reptiles in the colder, northern climates are viviparous.

Common coniferous forest reptiles Garter snakes, rattlesnakes, western pond turtles, and skinks are all found in the California redwood forests. Skinks are members of the lizard family, some of which climb trees. Some species are herbivorous, while others eat mostly insects.

African pine forests are home to the agama lizard. Other reptiles found in dry regions include many species of snakes such as the night adder, the puff adder, and the Gabonan adder.

BIRDS

Coniferous forests are home to hundreds of bird species. Some, such as insect-eating wood warblers, Canada geese, and northern goshawks, are migratory, which means they travel from one seasonal breeding place to another. Canada geese and northern goshawks return to northern forests by the thousands each spring and summer from areas with warmer winter climates.

During excessively cold or dry periods, birds can simply fly to more comfortable regions. Some, however, such as the blue jay, live in a particular forest year-round. If food is plentiful, the evening grosbeak, pine siskin, and red crossbill will also brave the winter weather.

Feathers protect birds not only from cold winters but also from tropical heat. Air trapped between layers of feathers acts as insulation.

Food Millions of insects attract summer birds, such as kinglets, woodpeckers, and flycatchers, to coniferous forests. Birds that stay in the forest all year must work hard for their food during the winter. For example, the nutcracker uses its powerful jaws to rip open pinecones for seeds. Blue grouse and the capercaillies, the largest species of European grouse, have an easier time of it, preferring to eat the conifer needles. Small birds and rodents are food for predators like eagles, owls, and hawks.

Reproduction All birds reproduce by laying eggs. One parent sits on the eggs to protect them from heat or cold until they hatch.

Common coniferous forest birds In temperate forests, common birds include screech owls, great horned owls, hummingbirds, woodpeckers, nuthatches, woodthrushes, American redstarts, hawks, blue jays, cardinals, scarlet tanagers, chickadees, and turkey vultures. Wrens, falcons, weaverbirds, thrushes, and chats are found in dry regions. Other common forest birds include the American dipper and the crossbill.

AMERICAN DIPPER The American dipper lives all year in mountain forests along the Pacific coast, from Alaska to Mexico. These birds stay near swift streams in which they wade and swim in search of insects, freshwater shrimp, snails, and fish. Since the dipper must dive into the water to catch its food, it must maintain a waterproof covering. The dipper spreads oil over its feathers from oil glands, and a special flap keeps water out of its nostrils. Its long toes help it grasp underwater surfaces, and it can stay submerged for about 30 seconds and dive as deep as 30 feet (9 meters).

RARE BIRD IN A RARE FOREST

Kirtland's warblers are rare songbirds that breed only in young jack pine forests in north-central Michigan. They usually build their nests on the ground under the small, growing trees. After jack pines reach more than 20 feet (6 meters) in height and no longer provide enough shelter, the birds abandon them. However, new jack pines will sprout and grow only after a forest fire, many of which are put out by humans. As a result, few new trees grow and, therefore, few warblers can raise families.

Dippers breed in February or March and lay their eggs in nests made out of materials found along the streams in which they hunt for food. Baby birds hatch after about two weeks and learn to dive very quickly.

CROSSBILL The crossbill lives year-round in coniferous forests and has a beak specially designed for digging the seeds out of pinecones. Only 5.5 to 6 inches (14 to 15 centimeters) in length, crossbills eat about 1,000 seeds each day. Different species of crossbills vary in color, like red or yellow, and their beaks are shaped differently depending upon where they live and what kind of pine cones they must open. The birds summer in the more northern forests in Alaska and Canada and winter in the warmer mountain forests of North Carolina and Oregon.

MAMMALS

Mammals are warm-blooded vertebrates covered with at least some hair, and they bear live young and produce their own milk. A few large mammals, such as mountain lions, bears, and deer, live in northern coniferous forests. Most coniferous forests are home to many small mammals, including mice, squirrels, woodchucks, and foxes.

During cold winters, mammals may burrow underground or find shelter among thick evergreens or under the snow. Snow dens are used by shrews and voles that remain active all winter. Some mammals, such as black bears, hibernate. During hibernation, the bear's heart rate drops from 25 to 10 beats per minute, which conserves body energy. Chipmunks accumulate deposits of fat below their skin that help insulate them and provide nourishment while they hibernate. In dry, warm climates, small mammals may remain inactive during hot weather.

Food Mammals may be plant eating or meat eating or both. Moss and lichens provide food for caribou, while downed trees and brush feed beaver and elk. Green plants, such as rushes, feed marmots and voles. Shrews eat insects that they dig out of the ground, while some carnivores like the wolf hunt in packs for deer and other animals. Bears are omnivores, which means that they eat both plants and animals.

Reproduction Mammals give birth to live young that have developed inside the mother's body. Some mammals, like the hare, are helpless at birth, while others, such as deer, are able to walk and even run almost immediately.

Common coniferous forest mammals Common mammals found in the boreal forest are wolverines, grizzly bears, lynxes, pine martins, minks, ermines, and sables. These animals are all predators and feed on other common boreal mammals like voles, snowshoe hares, marmots, and tiny tree mice. Chamois, red deer, moose, and elks live in the European mountain forests, while the mountain forests in Canada are home to moose, beavers, Canadian lynxes, black bears, and wolves.

TREE VOLE The tree vole spends most of its life at the top of fir trees in the Pacific Northwest and almost never walks on the ground. This small mammal is a type of rodent with a stout body and short tail. Voles construct treetop nests and tunnels from twigs and eat parts of pine needles.

WOLVERINE The wolverine is the largest member of the weasel family and grows to be about 4 feet (120 centimeters) long. This stocky carnivore can weigh up to 70 pounds (32 kilograms). Thick fur protects them from the snow and cold. Wolverines live in North America and Eurasia and are famous as savage hunters and ravenous eaters, capable of killing and eating an entire deer. They also feed on the marmot, beaver, hare, and fox.

Female wolverines give birth in early spring to two to four pups. The nest is usually in a crevice or other protected spot.

CHAMOIS The chamois (SHAM-ee) is a small goatlike antelope found in European mountain forests in the Alps, Pyrenees, and Dolomites. Living up to 25 years, an adult chamois weighs about 100 pounds (45 kilograms) and

A wolverine, the largest member of the weasel family, searching for food. (Reproduced by permission of Fieldmark Publications. Photograph by Robert J. Huffman.)

measures from 27 to 31 inches (69 to 79 centimeters) in length. Both males and females grow straight horns with backward-bending tips. In summer the chamois is reddish brown in color with a black stripe on the rump. In winter its coat turns black or brown, which helps to absorb heat.

The chamois has adapted to the mountains by developing rubbery hoof pads that help the animal keep its footing on slippery rocks. It eats mountain vegetation, especially clover, and some fir needles. If winter is especially harsh, the chamois moves to deciduous forests at lower elevations to find food.

GRIZZLY BEAR The grizzly bear is one of the fiercest animals in North America—strong enough to carry off full-grown cattle. Most have broad heads, extended jaws, big paws, and powerful claws. Grizzlies eat insects, like ants and bees, as well as seeds, roots, nuts, and berries. They also eat salmon, and are famous for their fishing skills.

ENDANGERED SPECIES

In the United States, acid rain (a mixture of water vapor and polluting compounds in the atmosphere that falls to Earth as rain or snow) has endangered the peregrine falcon. In Asia, South America, and Africa, all the big cats, including tigers, leopards, and cheetahs, are endangered. In the United States and Canada, the grizzly bear, beaver, and timber wolf is threatened, and in Australia, some species of kangaroos are threatened.

Grizzly bear Before the great expansion of the population westward, it is estimated that 100,000 grizzly bears lived in North America. By the 1990s, however, fewer than 1,000 existed, most of which lived in preserves such as Yellowstone National Park. Even though some are occasionally shot, the greatest problem for the bear's survival is the destruction of its habitat. Since bears are huge animals, weighing as much as 1,000 pounds (454 kilograms), they require large spaces to roam and huge quantities of food.

Beaver The beaver once ranged over North America from Mexico to the Arctic regions. It was so widely hunted for its fur and for a liquid called castorium, produced in the beaver's musk glands and used in perfume, that its numbers were reduced. The beaver is now confined largely to northern wooded regions. Beavers were also once common throughout northern Europe, where they are now virtually extinct, except in some parts of Scandinavia, Germany, and Siberia.

HUMAN LIFE

Without forests, there would probably be no human life on Earth. Many animals, including humans, are creatures of the forest. In early North America, native tribes, such as the Nootka and Haida, lived in the forests, hunting, trapping, and gathering for their survival.

IMPACT OF THE CONIFEROUS FOREST ON HUMAN LIFE

Forests have an important impact on the environment as a whole by contributing to the environmental cycles. From the earliest times, forests have also offered food and shelter, a place to hide from predators, and many useful products. Forests are an integral part of life on Earth.

Environmental cycles Trees, soil, animals, and other plants all interact to create a balance in the environment from which humans benefit. This balance is often maintained in cycles.

THE OXYGEN CYCLE Plants and animals take in oxygen from the air and use it for their life processes. When animals and humans breathe, the oxygen they inhale is converted to carbon dioxide, which they exhale. This oxygen must be replaced, or life could not continue. Trees help replace oxygen during photosynthesis, when they release oxygen into the atmosphere through their leaves.

A grizzly bear and her three cubs in Yellowstone National Park. Besides being one of the fiercest animals in North America, the grizzly bear is also one of the most endangered. (Reproduced by permission of the National Park Service. Harpers Ferry Center.)

THE CARBON CYCLE Carbon dioxide is also necessary to life, but too much is harmful. During photosynthesis, trees and other plants pull carbon dioxide from the air. By doing so, they help maintain the oxygen/carbon dioxide balance in the atmosphere.

When trees die, the carbon in their tissues is returned to the soil. Decaying trees become part of the Earth's crust, and after millions of years, this carbon is converted into oil and natural gas.

THE WATER CYCLE Coniferous forests shade the snow, allowing it to remain in deep drifts. The trees' root systems and fallen needles help build an absorbent covering on the forest floor, letting water from rain and melting snow trickle down into the Earth to feed underground streams and groundwater.

Not only do forests help preserve water in this way, but they also protect the land from erosion during heavy rain by acting as a wall or barrier. When forests are cut down there are no trees to act as barriers, and no tree roots to hold the soil in place, so it washes away. As a result, flooding is more common. For example, since 1997, parts of India and Bangladesh have suffered severe flooding caused in part by cutting of forests in the nearby Himalaya Mountains.

Trees take up water through their roots and use it for their own life processes. Extra moisture is then released through their leaves back into the atmosphere, helping to form clouds and continue the water cycle.

THE NUTRIENT CYCLE Trees get the mineral nutrients they need from the soil. Dissolved minerals are absorbed from the soil by the tree's roots and are sent upward throughout the tree. These mineral nutrients are used by the tree much like humans take vitamins. When the tree dies, these nutrients, which are still contained within parts of the tree, decompose and are returned back into the soil making them available for other plants and animals to use.

FOOD

Since the earliest times, forests have been home to game animals, such as rabbits, which have supplied meat for hunters and their families. Forests also supply fruits, seeds, berries, and nuts. Several species of southern conifers, for example, produce edible pine nuts, which are still collected by hand.

SHELTER

During prehistoric times, humans lived in the forest because it offered protection from predators and the weather. Today, people who choose to live in forested areas usually do so because they enjoy their beauty.

ECONOMIC VALUES

Forests are important to the world economy, since many products used commercially, such as wood, medicine, and resins and oils, are obtained from forests.

Wood Trees produce one of two general types of wood, hardwood or softwood, based on the trees' cell wall structure. Hardwoods are usually produced by deciduous trees, such as oaks and elms, while most coniferous trees produce softwoods. However, these names can be confusing, because some softwood trees, such as the yew, produce woods that are harder than many hardwoods and some hardwoods, such as balsa, are softer than most softwoods.

Wood is used not only for fuel, but also for building structures and manufacturing other products, such as furniture and paper. Hardwood from deciduous trees is more expensive because the trees grow more slowly. As a result, it is used primarily for fine furniture and paneling. Wood used for general construction is usually softwood such as white pine, Douglas fir, and spruce. Araucaria and kauri are commercially important conifers of the Southern Hemisphere. In order to conserve trees and reduce costs, some manufacturers have created engineered wood, which is composed of particles of several types of wood, combined with strong glues and preservatives. Engineered woods are very strong and can be used for many construction needs.

Medicines Since the earliest times plants have been used for their healing properties, and many drug companies maintain large tracts of forest as part of their research programs. For example, at one time the yew was considered

COMMON SOFTWOOD TREES

Africa	Asia	Australia	Central America
Pencil cedar	Benguet pine	Celery top pine	Caribbean pitch pine
Podo	Chir	Kauri	Yellowwood
Radiata pine	Hemlock	Rimu	
Thuya	Himalayan silver fir		
Yellowwood	I-Ching pine		
	Indian juniper		
	Japanese fir		
	Pencil cedar		
	Sugi		

a worthless "weed" tree and was burned by loggers after clearing a section of forest. Then taxol, an anti-cancer drug, was discovered in the bark of the yew. However, it takes the bark from six trees to make enough medicine for one cancer patient. Fortunately, in 1994, a synthetic form of taxol was created in the laboratory and the yews were spared.

Resins and oils Tree resins (REH-zihns; sap) and oils are also valuable. For example, conifer resins are used to make turpentine, paints, and varnish, while their oils are used as the fragrance in air fresheners, disinfectants, and cosmetics.

Recreation More people live in cities today than ever before, and many feel the need to occasionally escape to more natural surroundings. The beauty and quiet of coniferous forests draw many visitors for hiking, horseback riding, skiing, fishing, hunting, bird watching, or just sitting and listening to the sounds of nature.

Other resources Forest rivers are often dammed to provide a source of water for hydroelectric power. Norway, Sweden, Canada, and Switzerland rely heavily on hydroelectricity. The United States, Russia, China, India, and Brazil also use hydroelectric power, but on a smaller scale.

IMPACT OF HUMAN LIFE ON THE CONIFEROUS FOREST

About one-third of all the Earth's forests have been destroyed, and they continue to be lost at the rate of 6.75 million acres (2.7 million hectares) per decade. In North America alone, 90 percent of the Pacific Northwest forests are gone. Although it was once thought that forests of the far north were so

---

COMMON SOFTWOOD TREES

Europe	North America	South America
Austrian pine	Balasam fir	Alerce
Cedar	Cedar	Manio
European larch	Douglas fir	Monkey-puzzle tree (Chile pine)
Maritime pine	Larch	Parana pine
Norway spruce	Norway spruce	
Siberian yellow pine	Pine	
Silver fir	Sequoia	
Southern cypress		

inhospitable they were safe from human interference, the building of rail-roads made the area accessible. As a result, each year more land is being cleared for logging and mining operations.

Use of plants and animals As more and more forest land is being developed, native vegetation and wildlife habitats are destroyed. Trees are cut down and used for lumber and other products. In Canada, for example, much of the timber in the southern forests is gone, and northern forests are now being invaded by logging companies.

Not only trees, but other forest plants are in danger from overharvesting. In northern forests, mosses from the forest floor are used as fuel by construction workers and are being removed in large quantities. Without the insulation provided by the moss, permafrost is melting and causing floods. As too much logging destroys mountain forests, villages are at risk from landslides and avalanches that were once held in check by the dense trees.

Clear-cutting disrupts wildlife habitats, and new roads give hunters better access to wildlife, which sometimes results in overhunting. In some Canadian areas, for example, grizzly bears and bighorn sheep are now easy targets. Despite conservation efforts, parks and other protected areas are not

---

TIMBER!!

Logging is the harvesting of trees, sawing them into logs, and transporting them to a sawmill. About one percent of the world's timber is cut down each year. Half of it is used for fuel, and the rest for wood products, paper, and packaging materials. Most trees used for paper products are raised on tree farms and do not come from wild forests.

In the nineteenth century, logging was done by men using hand axes and saws. This method took a lot of people and time. In the 1950s, power chainsaws were used, and, by the 1970s, a variety of machines had revolutionized the logging industry.

Because logging machines can cut trees no bigger than about 2 feet (0.6 meter) in diameter, large trees are still cut by hand. A wedge is chopped from the trunk on one side of the tree and a cut is made with a saw on the other side. This causes the tree to fall in the direction of the wedge. After the tree is down, the limbs are removed and it is cut into lengths that can be easily moved.

In general, trees are removed either selectively or by clear-cutting. With selective methods, only certain trees are cut from a stand (group of trees). With clear-cutting, all the trees from several acres (hectares) of land are removed. Ideally, trees of the same species are then replanted, but many times this is not the case.

In India, some logging is done by "girdling." In girdling, a circular cut made around the tree trunk prevents water or nutrients from being carried to the branches. Several years later, when the tree is dead, it is harvested.

---

necessarily safe from development. In British Columbia, for example, mining interests have gained access to once-protected areas in parks.

Quality of the environment The quality of the forest environment is threatened not only by the direct effects of logging, mining, and hydroelectric development, but also by pollution and visitors. Roads, drilling rigs, and pipelines, all a necessary part of mining, destroy wilderness and disrupt habitats. Mercury, a poisonous liquid metal used in gold mining operations, and waste from chemical and petrochemical plants contaminate forest water sources. In Lake Baikal in the Siberian taiga, for example, mining activities have destroyed 5,500 acres (2,000 hectares) of life on the lake bottom.

Millions of acres (hectares) of forests in industrialized Europe, North America, and China are dead or dying from pollution and acid rain. Acid rain is a type of air pollution that forms when industrial pollutants such as sulfur or nitrogen combine with moisture in the atmosphere to produce sulfuric or nitric acids. These acids can be carried long distances by the wind

A lone dead pine in a clear-cut area in the Targhee National Forest in Idaho. Clear-cutting disrupts habitats, which may result in the endangerment of plants and animals. (Reproduced by permission of Corbis. Photograph by Raymond Gehman.)

before they fall either as dry deposits or in the form of rain or snow. When acid rain is absorbed into the soil, it can destroy nutrients and make the soil too acidic to support some species of trees. In coniferous forests, acid rain causes needles to drop and their color to fade and turn brown. Forests in northern Europe, southern Canada, and the eastern United States have been damaged by acid rain.

A large increase in the number of tourists all over the world has also put a tremendous stress on the forest environment. Many "wilderness" areas are being developed, and fragile forests are endangered as tourists hike, bike, ski, and snowmobile over the vegetation and disturb wild animals with their noise.

Forest management The National Forest Service was established in the United States to protect forest resources. More than 190,000,000 acres (76,760,000 hectares) of land are now publicly owned. Most of this acreage is west of the Mississippi. Forests in the eastern half of the United States are usually managed by state programs.

Most of these forests consist of areas that can be used for logging and other commercial purposes. However, the rest of the land is kept for recreation and conservation.

Many other nations, including Great Britain, Japan, China, and India, have also established programs to conserve and replant forests.

NATIVE PEOPLES

By 1950, native peoples in industrialized parts of the world had abandoned most of their tribal lands and customs. Many went to live in cities. However, in remote regions, some people still live a traditional lifestyle. For example, some taiga peoples are found in small Siberian towns that are accessible via the Trans-Siberian Railway. Other peoples live in mountain forests, especially in the Alps and Himalayas.

The Evenki In the northern Asian taiga of Siberia, Mongolia, and China live the Evenki, or Reindeer Tungus. Traditionally, the Evenki were nomads, surviving by means of hunting or reindeer herding. Although small groups still live this way, their traditional way of life is being threatened as oil, coal, and gas are mined and their homelands are taken over.

Many Evenki were removed from their lands by the government and settled onto collective farms after the Russian Revolution in 1917, and about 50,000 still live in Russia and China. In 1930, the Evenki national district was created, providing a permanent home for the people. This district contains some tundra vegetation and is covered by larch forest, but the climate is severe, with long, cold winters. The livelihood of the Evenki is supplemented by fur farming, farming, and jobs in industry or government.

Along with other native groups in Russia and China, like the Oroqen, the Evenki are working to preserve their culture.

The Cree The Ouje-Bougoumou Cree are a native people from the James and Hudson Bay area of Quebec in Canada. Some Cree tribes, collectively known as the Cree nation, lived on the plains and cultivated maize (corn). Others lived in the forests and fished and hunted caribou, moose, bear, beaver, and hare.

The Cree believe they have a special relationship with nature: nature provides them with food and they take care of the land and hunt only what is necessary for life. However, beginning in 1920 when the first non-native miners came onto their land searching for gold and copper, the Cree were forced to relocate their villages. Although about 12,000 Cree are left in the nation, only 4,000 of them live in a traditional manner, fishing, hunting, trapping, and farming. However, clear-cutting of the boreal forests and hydroelectric development in the James Bay area continue to threaten their lifestyle.

The Cree nation has been fighting for many years to stop the spread of mining and forestry and has sued the Canadian government in an effort to slow down the encroaching companies.

The Pehuenche The Pehuenche live in the forests of southeastern Chile. To them, the coniferous monkey-puzzle tree is sacred, and they refuse to cut it down. Its seeds are ground and turned into flour. Even though the Chilean government outlawed cutting the trees, logging continues and endangers the existence of both the people and the tree.

A Cree women cleans a caribou hide in the snow at a Cree camp in Canada. (Reproduced by permission of Corbis. Photograph by Abbie Enock; Travel Ink.)

THE FOOD WEB

The transfer of energy from organism to organism forms a series called a food chain. All the possible feeding relationships that exist in a biome make up its food web. In the forest, as elsewhere, the food web consists of producers, consumers, and decomposers. An analysis of the food web shows how energy is transferred within a biome.

Green plants are the primary producers in the forest. They produce organic materials from inorganic chemicals and outside sources of energy, primarily the Sun. Trees and other plants turn energy into plant matter, such as pinecones, needles, and seeds.

Animals are consumers. Primary consumers are plant-eating animals like squirrels, voles, mice, and beetles. Secondary consumers eat the plant-eaters. Tertiary consumers are the predators, like owls, wolves, and humans that eat other animals. Bears and humans are also omnivores, eating both plants and animals.

Decomposers eat the decaying matter from dead plants and animals and help return nutrients to the environment. Small underground insects called springtails and mollusks like the banana slug help this process of decomposition by breaking down dead plants. This allows other organisms, like bacteria and fungi, to reach the decaying matter and decompose it further.

SPOTLIGHT ON CONIFEROUS FORESTS

BONANZA CREEK EXPERIMENTAL FOREST

Bonanza Creek Experimental Forest is located south of Fairbanks, Alaska, in the Yukon-Tanana uplands in the Tanana Valley State Forest, south of the Brooks Mountain Range and north of the Alaska Range.

> **BONANZA CREEK EXPERIMENTAL FOREST**
> **Location:** Alaska
> **Area:** 12,487 acres (5,053 hectares)
> **Classification:** Boreal forest/taiga

Within the Yukon-Tanana uplands are forests, grasslands, wetlands, and alpine tundra (a cold, dry, windy region where no trees grow). Elevations in Bonanza Creek range from 393 to 1,542 feet (120 to 470 meters) above sea level. The Yukon and Tanana rivers cross the region.

The climate in the Alaskan taiga is extreme, with temperatures ranging from -58° to 95°F (-50° to +35°C). The average annual temperature is 26°F (-3.3°C). July is the warmest month with an average daily temperature of 61.5°F (16.4°C). January is the coldest with an average temperature of -12.8°F (-24.9°C). For about 233 days each year, freezing temperatures are possible. About 65 percent of the area's precipitation falls as rain in the summer; the other 35 percent is snow. The annual average precipitation is about 10.6 inches (269 millimeters).

Taiga soils are often poorly developed. A layer of permafrost (permanently frozen soil) is on the lowlands and northern slopes of mountains and prevents rain and melting snow from being absorbed into the ground. Since the moisture stays close to the surface, the soil is soggy.

Predominant trees in the area are white spruce, black spruce, paper birch, and aspen.

Numerous insects found on the taiga include budworms, sawflies, midges, and beetles. The northern wood frog is one of the few amphibians that live here.

Woodpeckers, swallows, sparrows, peregrine falcons, great horned owls, snowy owls, and boreal owls are among the birds that make the Bonanza Creek Forest their home.

Mammals include a number of voles and lemmings, muskrats, red squirrels, and meadow jumping mice. Moose and caribou migrate here to breed. Carnivores include coyotes, gray wolves, and wolverines.

OLYMPIC NATIONAL PARK

Olympic National Park is located on the Olympic Peninsula in the northwestern corner of Washington State. Forests in the park lie between sea level and 2,000 feet (609 meters) above sea level.

Among the coniferous forests in the park is a temperate rain forest found in its lower western and southern regions. Temperate rain forests grow at higher latitudes than tropical rain forests. They often occur along coastlines and experience cool winters and heavy year-round rainfall. The rain forest here receives about 150 inches (381 centimeters) of rain annually. Forests in the northern and eastern portion of the park are drier, getting about 60 inches (152 centimeters) of rain per year. The nearness of the ocean keeps the temperatures somewhat moderate with very few days of below-freezing weather in winter. In summer, temperatures rarely exceed 85°F (29°C).

> **OLYMPIC NATIONAL PARK**
> **Location:** Washington
> **Area:** 908,447 acres (363,379 hectares)
> **Classification:** Temperate evergreen forest

The forest floor supports toadstools, creepers, and ferns, such as sword, bracken, and licorice ferns. Club moss and algae hang from the trees. Shrubs found in the park include red huckleberry and salmonberry. The western thimbleberry can be recognized by its large, hairy, maple-like leaves. Wildflowers include western starflower, western trillium, and foamflower. Living deep in the woods is the deerfoot vanillaleaf.

Douglas fir and Pacific silver fir are common trees. The species of Sitka spruce found here rarely grows far from salt water. The grand fir also grows only in this region.

Invertebrates include the unusual banana slug, a scavenger and decomposer that helps to keep the forest floor and soil healthy. Springtails, golden buprested beetles, and questing spiders can also be found here.

The Pacific tree frog and the bullfrog live in the park, as do several species of snakes.

More than 200 species of birds are common to the Olympic Peninsula including flicker, crow, cliff swallow, winter wren, water ouzel, pileated woodpecker, northern spotted owl, winter wren, raven, and jay. Both golden and bald eagles also frequent the area. Ruffled grouse stay all winter and grow long, feathery "snowshoes" on their toes.

The park is home to one of the largest herds of Roosevelt elk, estimated at between 5,000 and 7,000 animals. An adult male can weigh as much as 600 pounds (299 kilograms), and the elk are a major hunting and tourist attraction. Other mammals found on the peninsula are the coyote, mountain beaver, Olympic marmot, raccoon, skunk, Columbian black-tailed deer, Rocky Mountain goat, black bear, mountain cottontail rabbit, and snowshoe hare. Cougars may be found in remote areas.

YELLOWSTONE NATIONAL PARK

The first national park in the United States, Yellowstone is located primarily in the state of Wyoming, with small portions in southern Montana and eastern Idaho. Mountains, lakes, rivers, and waterfalls cross the terrain. Most of the park is coniferous forest, but Yellowstone is also famous for the geyser (GY-zuhr) called Old Faithful, and its hot springs, steam vents, and mud caldrons.

Located on a high plateau, Yellowstone is 11,358 feet (3,462 meters) above sea level at its highest point. Its lowest elevation is 5,282 feet (1,610 meters). Most conifers are found between 7,000 and 8,500 feet (2,133 and 2,590 meters).

Annual precipitation varies from one end of the park to the other. In the north the average is 10 inches (26 centimeters). In the south, it is 80 inches (205 centimeters). Average annual snowfall is about 150 inches (380 centimeters). The average January temperature is 10°F (-12°C) and, in July, the temperature ranges between 55° and 70°F (13° to 25°C).

> **YELLOWSTONE NATIONAL PARK**
> **Location:** Wyoming
> **Area:** 2,219,791 acres (898,318 hectares)
> **Classification:** Mountain coniferous forest

Several species of conifers grow in the park, with lodgepole pine being the most common. Smaller stands of Douglas fir are found at lower elevations and, at higher elevations, Engelmann spruce and alpine fir abound. Also found here are more than 1,000 species of colorful flowering plants, including Indian paintbrush, columbine, and harebell.

The park supports about ten species of reptiles and amphibians. Its lakes are stocked with fish, many of which are non-native species.

Hundreds of species of birds live in the park, including the western tanager, goshawk, bald eagle, and dipper.

Large mammals include buffalo, elk, bighorn sheep, moose, and grizzly bear. Wolves, which had been eliminated by ranchers who feared they would prey on cattle, were reintroduced to the park in 1995. The park also supports a large herd of wapiti deer, or American elk. This once common animal is now found only in the Rocky Mountains and southern Canada. The wapiti is hunted for its hide, flesh, and head, which is usually displayed as a trophy.

CONIFEROUS FORESTS OF JAPAN

Japan is a chain of four islands in the Pacific Ocean—Hokkaido, Honshu, Shikoku, and Kyushu. The coniferous forests in Japan are found primarily in the mountains in the northern half of Hokkaido.

Japan's climate is influenced by the presence of mountain ranges and the Sea of Japan and varies from one region to another. In general, Japan receives more than 40 inches (1,020 millimeters) of precipitation annually, mostly as rain during the summer. The summer rainy season is in June and July, and typhoons (ocean storms) are not uncommon during this time. The southern islands are generally warmer. In Hokkaido, the average temperature in January, the coldest month, is 16°F (-9°C), and in August, the hottest month, it is 70°F (21°C).

> ### CONIFEROUS FORESTS OF JAPAN
> **Location:** Japan
> **Classification:** Mountain coniferous forest and temperate evergreen forest

Coniferous trees are found from sea level up to about 5,000 feet (1,525 meters) in the mountains. Common conifers in the forests of Hokkaido are Hondo and Sakhalin spruce and Maries and Veitch firs. Mosses and lichens are found on the forest floor and hanging from tree branches.

Many native forests have been destroyed and replanted with coniferous forests, which are managed as commercial plantations. Two of the most important timber trees are the hiba and the Japanese cedar. The cedars, often exceeding 150 feet (45.7 meters), have long trunks and reddish bark.

Natural stands of Japanese cedars, with trees more than 2,000 years old, cover about 4,133 square miles (10,747 square kilometers) on Yaku Island south of Kyushu. These trees grow in rocky areas with little soil and light, and as a result, their growth is stunted. Because the grain of their wood is tightly compacted and contains large quantities of resin, they do not easily decay. Many cedars on the island exceed the average life span of 500 years. The Yaku sugi tree, which also grows here, is a decay-resistant conifer related to the redwood. Also on Yaku are mixed forests containing broad-leaved trees. In 1993 the Yaku forests were declared a World Heritage Property by the World Heritage Convention.

Japan's forests are home to about 150 species of songbirds, as well as birds of prey such as eagles, hawks, and falcons. Bramblings and Eurasian nutcrackers are also common.

Brown bears, wild boars, Siberian chipmunks, Asiatic pika, and Hokkaido squirrels live in the forests. An unusual inhabitant is the raccoon dog, or tanuki, a native of eastern Asia that some scientists place in the dog family and others in the raccoon family. The tanuki looks like a raccoon, with dark face markings that stand out against a yellow-brown coat. Its long fur is sold commercially.

LAPLAND NATIONAL PARKS

Much of Scandinavia—Norway, Sweden, and Finland—is covered in coniferous forest. These boreal forests are interspersed with wetlands and lakes. Among the first to practice conservation methods, the countries of Scandinavia preserved much of their land and wildlife by setting aside national parks, nature parks, natural monuments, game reserves, and nature reserves. Despite logging, untouched forests are still found in Sweden and Finland.

> **LAPLAND NATIONAL PARKS**
> **Location:** Sweden
> **Area:** 1,292,300 acres (523,200 hectares)
> **Classification:** Boreal forest/taiga

The three Lapland National Parks share a common boundary. Sarek, with the most wilderness, is mostly alpine landscape with glaciers, tall mountain peaks, and alpine tundra (a cold, treeless region). Sjofallet is mountainous, but much of its original landscape was destroyed by the construction of a hydroelectric power facility. The coniferous forest is found in Padjelanta, an area with mountain peaks, large lakes, and some virgin forest.

Rare wildflowers dot the area and lichens are common in the open woodlands. Scots pine and Norway spruce are the most common conifers.

The wetlands and lakes produce much insect life, which attracts millions of waterfowl and wading birds in summer. The area is one of the few remaining nesting sites for the red-throated loon and the whooper, a noisy member of the swan family. The peregrine falcon and osprey also live here, as do the brown bear, lynx, wolverine, and moose.

Many people visit Padjelanta, which has marked trails and cross-country ski paths. Scientists visit to conduct research on plants, animals, geology, glaciers, and water resources. The parks are also used by the Lapps, a native people who keep thousands of reindeer here during the summer. Conservationists, however, protest the Lapps' use of motor bikes to herd the reindeer because the bikes damage the fragile environment. The Lapps also have rights to commercial hunting of grouse and ptarmigan.

The greatest danger to Scandinavian plant and animal life are acid rain and airborne pollutants that blow in from Europe and Russia.

CONIFEROUS FORESTS OF NEW ZEALAND

New Zealand is an island nation in the South Pacific, about 1,000 miles (1,600 kilometers) southeast of Australia. A temperate evergreen forest with mixed deciduous and coniferous trees grows in the north.

The climate of New Zealand has no real extremes. In general, summer temperatures are above 70°F (21°C), and winter temperatures are rarely below 50°F (10°C). Annual rainfall ranges from 12 inches (300 millimeters) to as much as 315 inches (800 centimeters), depending upon location. The average for the country as a whole is between 25 and 60 inches (127 to 152.4 centimeters).

> **CONIFEROUS FORESTS OF NEW ZEALAND**
>
> **Location:** New Zealand
>
> **Classification:** Southern Hemisphere forest

Conifers make up most of the canopy in the forests and include the huge kauri tree, as well as trees in the plum pine family. The rimu, or New Zealand red pine, grows as tall as 150 feet (45 meters) and its wood is reddish to yellowish brown. Rimu is used in construction and furniture making, and its bark contains a tanning agent that colors leather red. The radiata pine, a native of California, was transplanted in New Zealand and has become a key species there. Only a few flowers, like yellow kowhai and pohutakawa, grow in the New Zealand forests.

Very few animals are native to New Zealand. They include several species of frogs and bats, as well as two lizards, the gecko and the tuatara. Europeans introduced deer, opossums, and goats.

CONIFEROUS FORESTS OF RUSSIA AND SIBERIA

Almost one third of the world's forest resources are in Russia, where the boreal forest is the largest forest in the world and spans the region across the top of the Asian continent.

> **CONIFEROUS FORESTS OF RUSSIA**
>
> **Location:** Russia and Siberia
>
> **Classification:** Boreal forest/taiga

The climate in the forests varies. In the more northern section, the weather is harsh and summers are short. In the Yakut taiga, which is located in eastern Siberia, for example, winter temperatures may reach -85°F (-65°C). The average annual temperature is only around 10°F (-12°C). Temperatures in the middle taiga, such as that near Lake Baikal, are less severe, and conditions for tree growth are better. Here, the average annual temperature is about 19°F (-7°C), and the growing season lasts from 12 to 17 weeks. An even milder climate is characteristic in the southernmost region of the taiga where the growing season may last more than 20 weeks.

Conifers found here include pine, spruce, fir, larch, cedar, Japanese stone pine, Jeddo spruce, Asian white birch, and dwarf mountain pine. Many trees are stunted by the cold wind and, as a result, reach only about 19 feet (6 meters) in height. Where trees are protected from the wind, they grow about 50 feet (15.2 meters) tall. Shrubs like crowberry and bilberry grow in the understory. Mosses and lichens grow on the forest floor.

Some areas of the taiga are richer in wildlife than others. The more southern regions support hundreds of species of birds and about 50 species of mammals. The hazel hen and Siberian jay are unique to the taiga. Mammals include the moose, lynx, brown bear, Siberian red deer, wolverine, Asiatic chipmunk, northern pika, and sable. The sable is hunted for its highly prized fur, which has been called "soft gold." Once endangered, it is now being protected in nature reserves such as the Barguzin Nature Reserve near Lake Baikal.

CONIFEROUS FORESTS OF THE SOUTHEASTERN UNITED STATES

Temperate pinelands in the United States are found primarily in New Jersey, Virginia, North Carolina, South Carolina, Georgia, Alabama, Mississippi, Louisiana, Arkansas, Florida, and Tennessee. The pines that grow here are called southern pines, and loblolly, slush, longleaf, and shortleaf pines predominate. Many of these trees took over farms and plantations abandoned after the Civil War (1861–65).

> **CONIFEROUS FORESTS OF THE SOUTHEASTERN UNITED STATES**
> **Location:** Southeastern states
> **Classification:** Temperate pineland

The temperate climate with its mild winters favors tree growth. In Georgia, for example, average annual temperatures are about 40°F (4.4°C) in the mountains and 54°F (12 °C) on the southern coast. Average annual rainfall is about 50 inches (127 centimeters).

The soil of these forests tends to be rich and supports much tree growth, including oak and hickory, in addition to the many pines.

Wildlife includes many species of snakes and birds, as well as rabbits, squirrels, opossums, badgers, moles, deer, and wildcats.

FOR MORE INFORMATION

BOOKS

Burton, Robert, ed. *Nature's Last Strongholds.* New York: Oxford University Press, 1991.

Fitzharris, Tim. *Forests.* Canada: Stoddart Publishing Co., 1991.

Ganeri, Anita. *Habitats: Forests.* Austin, Texas: Raintree Steck-Vaughn, 1997.

Greenaway, Theresa, Christiane Gunzi and Barbara Taylor *Forest.* New York: DK Publishing, 1994.

Hirschi, Ron. *Save Our Forests*. New York: Delacorte Press, 1993.

Kaplan, Elizabeth. *Taiga*. Biomes of the World. New York: Benchmark Books, 1996.

Kaplan, Elizabeth. *Temperate Forest*. Biomes of the World. New York: Benchmark Books, 1996.

Pringle, Laurence. *Fire in the Forest: A Cycle of Growth and Renewal*. New York: Atheneum Books, 1995.

Sayre, April Pulley. *Taiga*. New York: Twenty-First Century Books, 1994.

Staub, Frank. *Yellowstone's Cycle of Fire*. Minneapolis, MN: Carolrhoda Books, 1993.

ORGANIZATIONS

Center for Environmental Education
  1725 De Sales Street NW, Suite 500
  Washington, DC 20036

Environmental Defense Fund
  257 Park Ave. South
  New York, NY 10010
  Phone: 800-684-3322; Fax: 212-505-2375
  Internet: http://www.edf.org

Environmental Network
  4618 Henry Street
  Pittsburgh, PA 15213
  Internet: http:/www.envirolink.org

Environmental Protection Agency
  401 M Street, SW
  Washington, DC 20460
  Phone: 202-260-2090
  Internet: http://www.epa.gov

Forest Watch
  The Wilderness Society
  900 17th st. NW
  Washington, DC 20006
  Phone: 202-833-2300; Fax: 202-429-3958
  Internet: http://www.wilderness.org

Friends of the Earth
  1025 Vermont Ave. NW, Ste. 300
  Washington, DC 20003
  Phone: 202-783-7400; Fax: 202-783-0444

Global ReLeaf
  American Forests
  PO Box 2000
  Washington, DC 20005
  Phone: 800-368-5748; Fax: 202-955-4588
  Internet: http://www.amfor.org

Greenpeace USA
  1436 U Street NW
  Washington, DC 20009
  Phone: 202-462-1177; Fax: 202-462-4507
  Internet: http://www.greenpeaceusa.org

Nature Conservancy
  1815 North Lynn Street
  Arlington, VA 22209
  Phone: 703-841-5300; Fax: 703-841-1283
  Internet: http://www.tnc.org

Sierra Club
  85 2nd Street, 2nd fl.
  San Francisco, CA 94105
  Phone: 415-977-5500; Fax: 415-977-5799
  Internet: http://www.sierraclub.org

World Wildlife Fund
  1250 24th Street NW
  Washington, DC 20037
  Phone: 202-293-4800; Fax: 202-293-9211
  Internet: http://www.wwf.org

WEBSITES

Note: Website addresses are frequently subject to change.

National Geographic Magazine: http://www.nationalgeographic.com

National Park Service: http://www.nps.gov

Scientific American Magazine: http://www.scientificamerican.com

Ouje-Bougoumou Cree Nation: http://www.ouje.ca

BIBLIOGRAPHY

Burnie, David. *Tree.* New York: Alfred A. Knopf, 1988.

Caras, Roger. *The Forest: A Dramatic Portrait of Life in the American Wild.* New York: Holt, Rinehart, Winston, 1979.

Davis, Stephen, ed. *Encyclopedia of Animals.* New York: St. Martin's Press, 1974.

Dixon, Dougal. *Forests.* New York: Franklin Watts, 1984.

Duffy, Eric. *The Forest World: The Ecology of the Temperate Woodlands.* New York: A&W Publishers, Inc., 1980.

"Forest." *Colliers Encyclopedia.* CD-ROM, P. F. Collier, 1996.

"Forestry and Wood Production; Forestry: Purposes and Techniques of Forest Management; Fire Prevention and Control." Britannica Online. http://www.eb.com: 180/cgi-bin/g?DocF=macro/5002/41/19.html

International Book of the Forest. New York: Simon and Schuster, 1981.

McCormick, Jack. *The Life of the Forest.* New York: McGraw-Hill, 1966.

Moore, David M., ed. *The Marshall Cavendish Illustrated Encyclopedia of Plants and Earth Sciences.* Vol. 7. New York: Marshall Cavendish, 1990.

Morgan, Sally. *Ecology and Environment.* New York: Oxford University Press, 1995.

Page, Jake. *Planet Earth: Forest.* Alexandria, VA: Time-Life Books, 1983.

Storer, John. *The Web of Life.* New York: New American Library, 1953.

Sutton, Ann and Myron Sutton. *Wildlife of the Forests.* New York: Harry N. Abrams, Inc., 1979.

CONTINENTAL MARGIN

The continental margin is that part of the ocean floor at the edges of the continents and major islands where, just beyond the shoreline it tapers gently into the deep sea. The continental margin is made up of the continental shelf, the continental slope, and the continental rise.

The continental shelf begins at the shoreline. It is flat and its width varies. For example, off the Arctic coast of Siberia it is 800 miles (1,280 kilometers) wide. However, there is no shelf off southeastern Florida. Rich sediment (particles of soil and decaying matter) from rivers that flow to the sea filters down to the shelf. Over time, deposits of these sediments may become many thousands of feet (meters) thick. At its deepest points, the continental shelf is usually less than 660 feet (200 meters) below sea level (the level surface of the sea). Although it is easier to explore than deeper areas of the ocean, there is still much to learn about it.

At the end of the continental shelf is a steep dip that marks the edge of the continent. This is called the continental slope, which descends to depths of 10,000 to 13,000 feet (3,048 to 3,962 meters) and ranges in width from 12 to 60 miles (20 to 100 kilometers). The continental slope usually resembles the edge of a mountain range, and in some places the drop is spectacular. For example, along the coast of Chile in South America, where the Andes Mountains meet the sea, the drop from the highest mountain peak on land—Aconcagua—to the bottom of the continental slope is more than 9 miles (14 kilometers).

Beyond the continental slope is the continental rise, where sediments drifting down from the continental shelf have collected. These deposits may extend as far as 600 miles (1,000 kilometers) out into the ocean floor where the deep-sea basin begins.

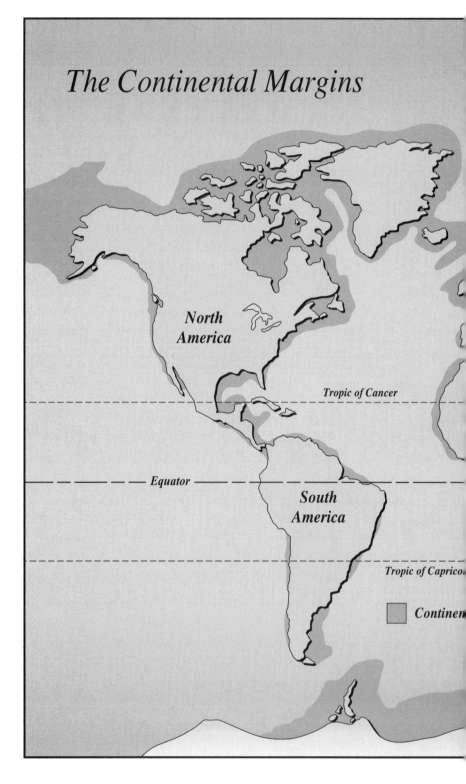

The Continental Margins

North America

Tropic of Cancer

Equator

South America

Tropic of Capricorn

Continen

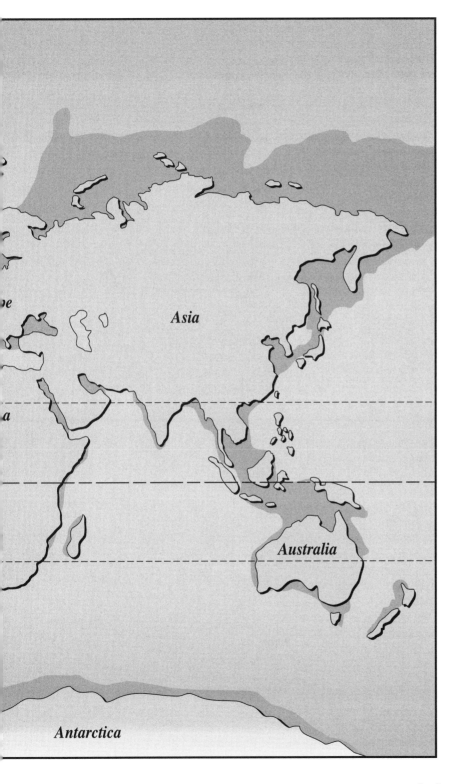

HOW THE CONTINENTAL MARGINS WERE FORMED

The World Ocean, which is all the oceans taken together, covers a total of 139,782,000 square miles (363,433,200 square kilometers)—about 70.8 percent of the Earth's surface. Over 200,000,000 years ago, the World Ocean was one body of water that surrounded one large continent. As time passed, this land mass began to pull apart. As a result, the continents and islands were formed.

The breakup of that one large continent was caused by heat forces welling up from deep within the Earth. As earthquakes split the ocean floor, molten rock from below the Earth's crust flowed into the fracture and became solid. For millions of years this process was repeated until the upper parts of the Earth's crust, on which the continents sit, were pushed even farther apart. About 50,000,000 years ago, the continents took their present shapes and positions.

About 20,000,000 years ago, when the sea level was at its lowest, the area which now makes up the continental shelves was above water. Forests

An illustration showing a cross-section of the continental margin, including the continental shelf and continental slope.

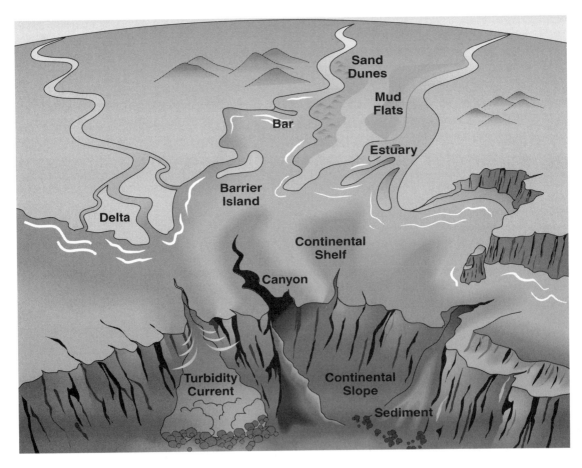

may have grown there, and it may have been home to many animals. Over millions of years, rain, wind, and wave action eroded (wore away) the shelf surface, and rivers and glaciers flowed across it. Gradually, sediments from the shelves were washed into the water. Later, as the glaciers melted, the sea level rose and covered the shelves so that the entire continental margin was under water.

THE WATER COLUMN

All of the waters of the ocean, exclusive of the sea bed or other landforms, is referred to as the water column. (For a more complete discussion of the water column, see the chapter titled "Ocean.")

Every element known on Earth can be found in ocean water. It is 3.5 percent dissolved salts by weight. The percentage of these salts determines the ocean's salinity (level of salts). These salts also make seawater heavier than fresh water. The ocean water closest to the surface is usually less salty because of rainfall and fresh water flowing in from rivers.

> ### SEA-GOING DINOSAURS
> About 63,000,000 years ago, a number of dinosaur species lived in the oceans. The Tylosaurus was 25 feet (7.6 meters) long and resembled a chubby crocodile with flippers rather than feet. Plesiosaurs, which could have been as long as 50 feet (15 meters), had flippers and long, giraffe-like necks. Flying reptiles, called Pteranodons, glided through the air over the ocean on leathery wings looking for fish. They returned to land only to lay their eggs.

The temperature of the oceans varies. In general, however, temperature changes are greatest near the surface where the heat of the Sun can be absorbed. In the warmest regions of the world, this heat absorption occurs to depths of 500 to 1,000 feet (150 to 305 meters).

ZONES IN THE OCEAN

Different parts of the ocean have different features, and different kinds of creatures live in them. These different parts are called zones. Some zones are determined by the amount of light that reaches them.

Over the continental shelves, the ocean receives enough light to support photosynthesis, the process by which plants use the energy from sunlight to change water and carbon dioxide into the sugars and starches they use for food. These surface waters, called the sunlit zone, reach down as far as 650 feet (198 meters) below the surface. The sunlit zone supports more plant and animal life than any other zone.

Below the sunlit zone and extending about halfway down the continental slope is the twilight zone, which ranges from 650 to 3,300 feet (198 to 1,006 meters) in depth. Only blue light can filter down to this level. It is too dark for plant life here, but animals can live at this depth.

Beginning about halfway down the continental slope and extending into the deepest region of the oceans is the dark zone. Like the twilight zone, the dark zone is unable to support plant life, but a variety of animals are able to live in its depths.

CIRCULATION

The oceans are constantly, restlessly moving. This movement takes the form of tides, waves, and currents, all of which affect the continental margins.

Tides Tides are rhythmic movements of the oceans that cause a change in the surface level of the water. They are created by a combination of the gravitational pull of the Sun and Moon and the Earth's rotation. High tide occurs when the water level rises. When the level lowers, it is called low tide.

Sea level refers to the average height of the ocean when it is halfway between high and low tides and all wave motion is smoothed out. Sea level changes over time.

Waves Waves are rhythmic rising and falling movements on the surface of the water. Most surface waves are caused by wind. Their size is due to the

Boats are left high and dry at low tide on a French coastline. (Reproduced by permission of Corbis. Photograph by Nik Wheeler.)

speed of the wind, the length of time it has been blowing, and the distance over which it has traveled. Breakers are waves that collapse on a shoreline because the water at the bottom of the wave is slowed by friction as it rolls along the shore. The top of the wave then outruns the bottom and topples over in a heap of bubbling foam.

One type of wave called a tsunami (soo-NAH-mee) is mainly caused by undersea earthquakes. When the ocean floor moves during the quake, its vibrations create a powerful wave that travels to the surface. When tsunamis strike inhabited coastal areas, they can destroy entire towns and kill many people.

Currents Currents are the flow of water in a certain direction. They can be both large and strong. Most currents are caused by the wind, the rotation of the Earth, and the position of continental landmasses. In the North Pacific, for example, currents moving west are deflected (pushed) northward by Asia and southward by Australia. The same currents then move east until they reach North and South America, which send them back toward the equator. Longshore currents are those that move along a shoreline.

The remains of houses in Hilo, Hawaii, after they were struck by a tsunami. Tsunamis are caused by undersea earthquakes and are capable of destroying entire towns. (Reproduced by permission of Corbis.)

Upward and downward movement of water also occurs in the ocean. Vertical currents are mainly caused by differences in water temperature and salinity. In some coastal areas, for example, strong wind-driven currents carry warm surface water away. Then an upwelling (rising) of cold water from the deep ocean occurs to fill the space. This is more common along the western sides of the continents. These upwellings bring many nutrients from the ocean floor to the surface waters.

At the continental margins, large quantities of sediments enter the ocean and move out along the seafloor. These sediments are often sped along by turbidity (tur-BID-ih-tee) currents. These currents may be caused by earthquakes or sudden slumping of loose sediments. The thick mixture of sediment and water rushes down the continental slope and through any submarine canyons at considerable speeds for long distances, much like an avalanche of snow. They are so strong, they have been known to break underwater cables that lay in their path hundreds of miles away. Turbidity currents can cause many changes in the margin floor.

GEOGRAPHY OF CONTINENTAL MARGINS

The general shape of the continental margin is usually determined by the shape of the coastline from which it extends. If it extends from a plane, then the margin will be broad and level. If it extends from a mountainous coast, then it will be steep and rocky. Steep cliffs that may have been formed by wave action when the level of the sea was lower may now be submerged and form part of the margin.

The present shape of a continental margin may also be due to other influences. Movement of the Earth's crust may have given it a "folded" appearance where the sea floor cracked and was pushed underneath the margin. Huge boulders and rocks may indicate that a glacier once moved across the region. The presence of a river may mean a larger quantity of sediment where it enters the ocean, and the weight of tons of this sediment may have forced the underlying rock to sink.

A continental margin located in an area where there are many earthquakes or volcanoes is considered an active margin because it is constantly changing due to the continuous earthquake and volcanic activity. Active margins are often found in the Pacific Ocean, such as along the west coast of South America. They are narrow and often drop sharply into a deep trench (steep valley).

Passive margins are quieter and usually free of earthquake and volcanic action. These are found where the ocean floor is still gradually spreading. Usually there is also a wide rise. Continental margins along the Atlantic and Indian Oceans are of this type.

DAMS

Volcanic action, earthquakes, or a reef (a ridge of rock or other material) may create a dam at the end of the slope or rise. This dam causes sediments to build up and become an extension of the shelf. A new slope and rise form at its end. Salt domes, huge mounds of salt that have moved upward from beds buried deep in the ocean floor, can also form dams and trap sediment. A salt dam has helped form the shelf in the Gulf of Mexico off the U. S. coast.

SUBDUCTION ZONES AND TRENCHES

Even today, the sea floor continues to spread. When it presses against the edges of the continents, they resist its movement. This results in an area of extreme pressure called a subduction zone. In the subduction zone, this enormous pressure forces the sea floor to crack, pushing it down and causing it to slide under the continental margin, often causing a deep, V-shaped trench to form. The greatest depths in the oceans are found in these trenches,

Salt domes, such as these, can help form continental shelves. (Reproduced by permission of Corbis. Photograph by Kevin Schafer.)

and the deepest trenches are located in the Pacific. The Mariana Trench is the deepest at 36,163 feet (11,022 meters). Many earthquakes occur in subduction zones.

UNDERWATER CANYONS

The continental slope is often cut by deep, V-shaped, underwater canyons (deep, narrow depressions in the Earth's crust). These canyons vary in size, although some can be very large, they are not as deep as trenches. A canyon off the coast of Africa, for example, extends for 15 miles (25 kilometers), but its depth is only 1,480 feet (450 meters).

The origin of canyons is also different from that of trenches. Some canyons appear to be related to rivers on land that during prehistoric times may have extended into the area now covered by ocean. The huge Monterey Canyon that lies off the coast of California is an example. The Salinas River once flowed into its bed during the Ice Ages.

Other canyons are more mysterious and no one is sure how they were created. The Bering Canyon north of the Aleutian Islands in the North Pacific Ocean is an example. It cuts through the continental margin for more than 680 miles (1,100 kilometers) and is the longest canyon in the world. However, there is no evidence it was ever associated with a river and no one knows why or how it was formed.

REEFS

A reef is a ridge or wall of material lying close to the surface of the water just offshore. Some reefs are made of rock and some of coral. A coral reef is a wall formed in shallow ocean areas by small, soft, jelly-like animals called corals. Corals attach themselves to hard surfaces and build a shell-like external skeleton. Many corals live together in colonies, the younger building their skeletons next to or on top of older skeletons. Gradually, over hundreds, thousands, or millions of years, a wall, or reef, of these skeletons is formed. Reef corals cannot live in cold waters and, as a result, are found only in warm waters north and south of the equator. The water must also be clear and free of sediment, and coral reefs do not form where rivers flow into the sea.

When a coral reef grows tall enough to break the surface of the water, it may begin to collect sand, gravel, weeds, and other matter. Gradually, an island, called a cay (KEE), begins to form. The wind or visiting animals bring seeds and, eventually, plants grow. As the plants die and decay, soil forms. After a time, the island may be able to support certain species of trees, such as mangroves.

Three main types of coral reefs are found on the continental margins: shelf reefs, fringing reefs, and barrier reefs. Shelf reefs form on a continental shelf having a hard, rocky bottom. A shallow body of water called a lagoon may be located between the shelf reef and the shore. Fringing reefs develop close to the land and no lagoon separates them from the shore. The longest is found in the Red Sea, where it stretches for 2,800 miles (4,500 kilometers). Barrier reefs line the edge of the continental shelf and separate it from deep ocean water. A barrier reef may enclose a lagoon and even small islands.

ESTUARIES AND DELTAS

When a river traveling through lowlands meets the ocean in a semi-enclosed channel (stream bed) or bay, the area is called an estuary. The water in an estuary is brackish—a mixture of fresh and salt water. In these gently sloping areas, river sediments collect creating muddy shores. When the muddy sediments form a triangular shape over the continental shelf, the area is called a delta.

GLACIAL MORAINES

A glacial moraine is a pile of rocks, gravel, and sand created when a glacier moves across the surface of the land. Glacial moraines are found on some continental shelves that were crossed by glaciers during the Ice Ages more than 10,000 years ago. When the glaciers retreated (shrank), the level of the sea rose and covered the area with water. The tops of some glacial moraines remain above the water and form islands. The Aland Islands in the Baltic Sea are an example.

An uninhabited sandy cay off Jost Van Dyke in the British Virgin Islands. Cays are formed from coral reefs that grow tall enough to break the surface of the water. (Reproduced by permission of Corbis. Photograph by Joel W. Rogers.)

BARS AND SHOALS

Where tides and currents can move large quantities of sediment, such as sand or gravel, bars and shoals may form. A bar is a ridge of sand that accumulates across a channel. Shoals are areas where enough sediment has accumulated that the water is very shallow and dangerous for navigation. The shoals around Nantucket Island, off Massachusetts, for example, have sunk a total of 2,100 ships.

PLANT LIFE

Most ocean plants live in the waters above the continental shelf. They include tiny, one-celled organisms and many kinds of seaweed and seagrasses. Some scientists estimate that the plants over the continental shelf produce more oxygen for Earth than all the forests on land.

Ocean plants are surrounded by water at all times. For this reason most have not developed the special tissues and organs needed by land plants for conserving water. Seaweeds, for example, use their "roots" only to anchor them in one spot, not to draw water from the soil.

The water also offers support to ocean plants. Even a small tree on land requires a tough, woody stem to hold it erect, but giant underwater plants do not require woody portions because the water helps hold them upright. As a result, their stems are soft and flexible, allowing them to move with the current without breaking.

Marine plants can be classified as either plankton or benthos. Plankton (a Greek word that means "wanderers") are plants that float freely on the water's surface. Benthic plants, however, anchor themselves in the sea floor. (Benthos in Greek means "sea floor.")

Ocean plants can be divided into two main groups: algae (AL-jee) and green plants.

ALGAE

Most ocean plants are algae. (Although it is generally recognized that algae do not fit neatly into the plant category, in this chapter we will discuss them as if they were plants.) Certain types of algae have the ability to make their own food by means of photosynthesis. Others, however, absorb nutrients from their surroundings.

WORLD'S BIGGEST ICE CUBES

Thousands of icebergs form each year and float freely in the ocean. Those in the Arctic are usually produced by large glaciers moving across Greenland. As the glacier reaches the ocean, huge portions break off. Arctic icebergs often contain soil and other evidence of their origin on land. Antarctic icebergs are gigantic chunks of pack ice, which form on open water. Unlike glacier ice, they do not originate on land, and therefore, contain no sediments.

The exposed portion of an iceberg may be more than 200 feet (61 meters) high, but this is only a hint of its actual size. About 90 percent of an iceberg remains underwater. That means its submerged portion could be 1,800 feet (549 meters) deep!

Some forms of algae are so tiny that they cannot be seen with the naked eye. They are called phytoplankton because they float freely in the water, allowing it to carry them from place to place. Other species of algae are massive and live in vast underwater forests anchored to the seafloor. These are the benthic species.

Growing season Algae contain chlorophyll, a green pigment used to turn energy from the Sun into food. As long as light is available, algae can grow. In some species, the green chlorophyll is masked by orange-colored pigments, giving the algae a red or brown color.

The growth of ocean plants is often seasonal. In some areas, such as the Arctic, the most growth occurs during the summer when the Sun is more nearly overhead. In temperate (moderate) zones, growth peaks in the spring but continues throughout the summer. In regions near the equator, no growth peaks occur since growth is steady throughout the year.

Food Most algae grow in the sunlit zone, where light is available for photosynthesis. Algae also require other nutrients that must be found in the water, such as nitrogen, phosphorus, and silicon. In certain regions, upwelling of deep ocean waters during different seasons brings more of these nutrients to the surface. This results in algal "blooms," sudden increases in the number of algae. Algal blooms also occur when nitrogen and phosphorus are added to a body of water by sewage or by runoffs from farmland.

Reproduction Algae reproduce in one of three ways. Some split into two or more parts, each part becoming a new, separate plant. Others form spores (single cells that have the ability to grow into a new organism). A few reproduce sexually, during which cells from two different plants unite to create a new plant.

Common algae Two types of algae commonly found along the continental margins include phytoplankton and kelp.

PHYTOPLANKTON Phytoplankton are microscopic one-celled plants that float on the surface of the water, always within the sunlit zone. They are responsible for about 90 percent of the photosynthesis carried out in the oceans. During photosynthesis, phytoplankton release oxygen into the atmosphere. Two forms of phytoplankton, diatoms and dinoflagellates (dee-noh-FLAJ-uh-lates), are the most common.

Diatoms have simple, geometric shapes and hard, glasslike cell walls. They live in colder regions and even within arctic ice. Dinoflagellates have two whiplike attachments that make a swirling motion. They live in tropical regions (regions around the equator).

KELP Kelp are a type of brown benthic algae that grow on rocks in temperate waters (41° to 72°F [5°C to 22°C]). In northern regions, benthic plants grow below 200 feet (61 meters). In tropical regions, where the Sun's rays are more vertical and can penetrate farther into the water, they grow as deep as 400 feet (122 meters). There is evidence that some can grow as deep as 1000 feet (305 meters), where no light penetrates, because they may obtain food from decomposing matter that filters down from above instead of by means of photosynthesis.

Although there are several different species of kelp, they take only two basic forms. One type has a simple trunk between 20 inches and 8 feet (50 centimeters and 2.5 meters) long and a leaflike frond (branches) on top. The second type may grow to more than 250 feet (76 meters) in length and has fronds all along its trunk. Some kelp have gas-filled chambers at the bases of their fronds, which help them remain upright. Although they resemble green plants, kelp have no true leaves, stems, or roots. Instead they have a rootlike structure, called a holdfast, that anchors them to the ocean floor.

Kelp reproduce sexually; cells from a male and female plant unite to form another plant. The giant kelp are the world's fastest growing plant, often growing as much as 20 inches (50 centimeters) in a single day.

Kelp grow in huge, floating groups, or forests, that cover hundreds of square miles (square kilometers) and provide shelter and food for animals such as fish, crabs, and sea otters. On the water's surface, the kelp can be so thick that sea otters can lie down on them as if in a hammock. Sea urchins, which also feed on kelp, can grow so numerous that they eventually destroy the kelp beds unless checked by natural predators.

GREEN PLANTS

The true green plants found in the ocean are seagrasses, such as wireweed and paddleweed, and are similar to land grasses. Unlike algae, they have roots and bloom underwater. Beds of seagrass occur in sandy bottoms in areas protected from currents, such as in lagoons or behind reefs. Large beds slow the movement of water and help prevent erosion of the shelf. Some marine animals use seagrasses for food and for hiding places.

ANIMAL LIFE

The oceans are the largest animal habitat on Earth, and most species are found along the conti-

---

TINY, BUT DEADLY

Pfiesteria piscidia, a species of tiny, one-celled dinoflagellates, has suddenly become more dangerous than its size would indicate. Beginning in the late 1980s, this normally nontoxic organism turned poisonous and began killing fish in North Carolina waters. Over a billion fish have been destroyed since 1991.

Pfiesteria release strong poisons into the water. The poisons cause fish to gasp for oxygen and develop bleeding sores on their bodies. *Pfiesteria* then feed on the dying tissues. The poisons are also harmful to shellfish and even mammals. This includes humans, who suffer impaired memory and learning abilities.

Most dinoflagellates obtain food by means of photosynthesis, but *Pfiesteria* passes through many different life stages, and in some of those stages, it appears to dine on other organisms. Experts suspect that the changes in *Pfiesteria* may have been caused by fertilizer and sewage runoff from hog and chicken farms in the area.

---

nental margins. Although the continental shelves underlie only about 7.5 percent of the total area of the oceans, their shallow waters support more forms of life than any other area in the ocean—perhaps any other place on Earth. Scientists estimate that as many as 30,000,000 species of sea life may still be undiscovered. The types of animals found along the upper continental slope depend on whether the floor is sandy, muddy, or rocky. The lower slope and rise are home to fewer animals.

Animals that live in the sea have developed ways to cope with its high salinity. Naked (covered with thin "skins" or shells) animals maintain high levels of salt in their blood normally and do not need to expel any excess. Others, such as most fish, have special organs that remove the extra salt from their system and release it into the water.

The water offers support to marine (ocean) animals as it does to plants. Many marine animals have special chambers in their bodies that allow them to adjust their buoyancy (BOY-un-see; ability to float) so that they can float in either shallow or deep water. Some, such as seals, have flippers to make swimming easier. Others, such as octopi, forcefully eject water in a kind of jet stream to help them move about.

In addition to being classified as microorganisms, invertebrates, or vertebrates, marine animals, like plants, can be classified according to their range and style of movement. Plankton drift in the currents and include such animals as jellyfish. Many plankton also move up and down the water column by regulating the amount of gas, oil, or salt within their bodies. (Production of gas or oil and removal of salt causes the organism to rise; the reverse causes them to sink.) Many larger animals spend part of their young lives as plankton. Crabs, which move about on legs as adults, but float with the current in their larval form, are an example. Larger animals that swim on their own, such as fish and dolphins, are called nekton. Benthos are the animals that live on the seafloor. These include snails and clams.

MICROORGANISMS

Most microorganisms are zooplankton (tiny animals that drift with the current). They include foraminifer, radiolarians, acantharians, and ciliates, as well as the larvae or hatchlings of animals that will grow much larger in their adult form. Some zooplankton eat phytoplankton and are preyed upon by other carnivorous (meat-eating) zooplankton, such as arrow worms.

Bacteria Bacteria are another type of microorganism found throughout the ocean. They make up much of the dissolved matter in the water column. As such, they provide food for lower animals. They also help decompose (decay) the dead bodies of larger organisms, and for that reason their numbers increase along the continental shelf where the most animal life is found.

INVERTEBRATES

Animals without backbones are called invertebrates. Many invertebrate species, such as worms and squids, are found in the ocean. Probably the most numerous and diverse group of invertebrates are crustaceans, which have hard outer shells for protection from potential predators. Crustaceans include animals such as lobsters and crabs.

Food Invertebrates may eat phytoplankton, zooplankton, or both. Some also eat plants or larger animals. The cone shell uses a harpoon-like tooth to spear its prey and inject it with poison. Others, including crayfish, roam along the bottom to feed on dead organisms by grasping them with their large pincher-like front claws.

Reproduction Marine invertebrates reproduce in one of three ways. (1) Eggs are laid and fertilized externally (outside the female's body), a parent watches over the young in the early stages, and offspring number in the hundreds. (2) Fertilization is internal (inside the female's body), a parent cares for the young in the early stages, and offspring number in the thousands. (3) Fertilization is external, the young are not cared for, and offspring number in the millions. Survival depends upon the absence of predators and the direction of currents.

Common invertebrates Invertebrates found along the continental margins range from planktonic jellyfish to nektonic octopi to benthic starfish and corals.

Jellyfish are commonly found close to shore where they float on the surface like balloons partly filled with air. All jellyfish move by squeezing their bodies to push out water that forms a kind of jet stream behind them. Most of the time, however, they live as plankton and simply float with the current. Jellyfish eat small prey, such as shrimp, tiny fish, and other plankters, which they catch by stinging them with their tentacles.

Another typical invertebrate that lives along the continental margins is the octopus. Octopi may crawl around the ocean bottom using their tentacles, or move through the water by means of jet propulsion. Different species are found throughout the ocean, and they vary in size. Those found in shallow water are generally the smallest, although giants with a tentacle spread of 32 feet (9.7 meters) have been found off the coast of Alaska.

REPTILES

Reptiles are cold-blooded vertebrates, which means their body temperature changes with the temperature of their surroundings. In cold temperatures they become sluggish but can still function. This means they do not have to waste energy keeping their body temperatures up as most mammals and birds must.

Only one species of lizard lives in the sea, the marine iguana. A saltwater crocodile is found in the waters of Southeast Asia, where it lives near the mouths of rivers. However, the reptiles most commonly found in the oceans are the sea turtle and the sea snake.

Food Sea turtles eat soft plant foods as well as small invertebrates, such as snails and worms. Turtles have no teeth. Instead, the sharp, horny edges of their jaws are used to shred the food enough so they can swallow it.

Sea snakes are carnivorous, feeding primarily on fish and eels, which they find along the ocean bottom in rocky crevices. They first bite their prey, injecting it with poison so that it cannot escape.

Reproduction Both snakes and turtles lay eggs. Turtles lay theirs in a hole which they dig with their flippers on a sandy shore. They then cover the eggs with sand and abandon them, taking no interest in the offspring. Six weeks later the eggs hatch and the young turtles make a run for the water and disappear.

Some species of sea snakes come to shore to breed and lay their eggs on land. Others bear live young at sea.

Common marine reptiles Sea turtles can be distinguished from land turtles by their paddle-like limbs called flippers, which enable them to swim. Sea turtles have glands around their eyes that remove excess salt from their bodies, a process that makes them appear to cry. At least one species of turtle hibernates on the seabed during the winter. Green turtles are migrators, traveling as far as 1,250 miles (2,000 kilometers) to return to a particular breeding area where they lay their eggs.

At least 15 species of sea snakes live in tropical oceans (those around the equator), and half of these are found in regions around Australia and New Guinea. They look much like land snakes, having long bodies, some of which attain 9 feet (2.7 meters) in length. They also have salt glands that help them maintain a body fluid balance. Their tails are paddle shaped, which helps them move through the water.

FISH

Fish are primarily cold-blooded vertebrates that have gills and fins. The gills are used to draw in water from which oxygen is extracted, and the fins help propel the fish through the water. Although most fish are long and sleek in design, shapes vary greatly. Manta rays, for example, are flat and round, while seahorses are narrow and swim in a vertical position.

> ### THE WORM THAT DOESN'T NEED A WATCH
>
> Some animals seem to have built-in "clocks," like one species of marine worm called the palolo worm. This little creature, which lives in the waters of the central Pacific, develops eggs and sperm in the rear half of its body. When the time is right, this rear section detaches from the rest of its body and swims to the surface. If that weren't strange enough, all this happens like clockwork on the day of the last quarter of the October-November Moon. On this day, the surface waters grow thick with the rising egg sections from millions of worms.

Most fish species live over the continental shelf in the sunlit zone. However, many species live in the twilight and dark zones, including the hatchetfish, the swallower, and the black dragonfish. Twilight and dark zone fish are discussed in more detail in the chapter titled "Ocean."

Food Some fish, such as anchovies, are plant eaters whose diets are primarily phytoplankton, algae, or sea grasses. Fish that must swim in search of prey, such as swordfish, have more streamlined bodies than those, like the sea robin, that glide close to the bottom sediments in search of a meal. Many dark zone species eat carrion (decaying flesh).

Reproduction Most fish lay eggs, although many bear live young, such as certain species of sharks. Some, such as the Atlantic herring, abandon the eggs once they are laid. Others build nests and care for the new offspring. Still others carry the eggs with them, usually in a special body cavity or in their mouths, until they hatch.

Certain fish, such as sturgeons, travel thousands of miles to return to a particular breeding area where they lay their eggs and then die. By some inherited means of guidance, their young will make the same journey when their turn comes to breed, and the cycle is repeated.

A DENTIST'S DREAM

Sharks may have as many as 3,000 teeth in their mouths at one time. These teeth are arranged in rows, and only the first few rows are used. If a tooth is lost during feeding, the next tooth in line moves up to take its place.

Common fish Fish are so numerous and so varied over the continental margins that few could be considered "typical." However, mackerels and rays are commonly found there.

Mackerels live over the continental shelves in temperate and tropical oceans. They gather in schools (groups) in the upper waters during the warm months and descend to deeper waters in the winter. Their primary food is zooplankton, such as fish eggs. Mackerels lay their eggs in midwater where they drift with the currents. Some species are popular as game fish, and many species are important commercially in the fishing industry.

Rays are relatives of the shark, but they have flat, broad bodies. Their eyes are on top, and their mouths are on the underside. They feed by flapping and gliding over the seabed in search of clams and other similar prey. Some species have the ability to give an electric shock for defense or to kill prey. The largest species, the devil or manta ray, can grow as much as 20 feet (6 meters) in diameter. Most species bear live young.

SEABIRDS

Most seabirds remain near land where they can nest during breeding season. Many have adapted to marine environments by means of webbed feet and special glands for removing excess salt from their blood.

Food All seabirds are carnivorous. Most eat fish, squid, or zooplankton and live where food is plentiful. Several species, such as sea ducks, dive to the bottom to feed on shrimps, worms, or crabs. Other species, such as cormorants, spot their prey from high above in the sky and then plunge in just deep enough to catch it. Some, like terns, swoop down on fish swimming close to the surface.

Reproduction Most seabirds nest on land. Some nest in huge colonies on the ground, others dig burrows, and still others prefer ledges on cliffs. Like land birds, seabirds lay eggs and remain on the nest until the young are able to leave on their own. Some live and feed in one area and migrate to another for breeding. Birds that nest on sandy shores tend to lay speckled or blotchy eggs in beige and brown colors that blend in with the sand and pebbles.

Common seabirds Birds that live along the ocean's margins include gulls, plovers, gannets, pelicans, boobies, frigate birds, puffins, and penguins. There are only a few true sea ducks, the eider and the scoter.

MARINE MAMMALS

Mammals (warm-blooded vertebrates that bear live young) must come to the surface to breathe, and many, such as seals and polar bears, live part of the time on land. Whales, porpoises, sea cows and their relatives remain in the water at all times.

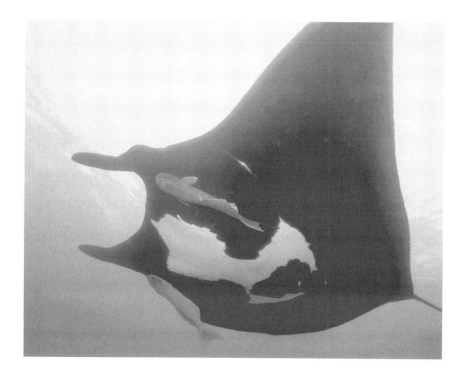

The largest species of ray, the manta ray, can grow to as much as 20 feet (6 meters) in diameter. (Reproduced by permission of Corbis. Photograph by Brandon D. Cole.)

Mammals must maintain a high body temperature. Most, therefore, have a special layer of fat below their skin that protects them from the cold water. Some, such as walruses, have very thick skins, which makes it difficult for them to keep cool in the Sun. Others, such as seals, are covered with fur that helps insulate them against the cold.

Food The dugong, a type of sea cow, is the only plant-eating mammal that truly lives in the sea. Most marine mammals are carnivorous. Some, like seals and walruses, feed on fish and squid that live on the continental shelf. Others, like killer whales, hunt for seals and other mammals.

Reproduction Most marine mammals have only one offspring at a time. The young are nursed on the mother's milk until they are able to find food on their own. This is true whether the mammal spends all its time in the water or part of the time on shore.

Common marine mammals Whales, dolphins, and porpoises are mammals that resemble fish. The killer whale, a relative of the dolphin, hunts in large groups and feeds on other mammals as well as fish and birds. Another fishlike mammal, the narwhal, or sea unicorn, has a tooth that grows into a spiral as much as 10 feet (3 meters) in length.

Seals are common along the continental shelves where they may plunge as deep as 3,300 feet (1,000 meters) in search of fish. Some species can stay underwater for as long as 30 minutes. Seals usually gather in huge on-shore colonies to breed.

Although polar bears are usually thought of as land animals, they spend much of their time in the water and may swim 20 miles (32 kilometers) or more in search of food. Their primary prey is seals. They may wait for hours along a shore for a seal to surface. If they catch it, they drag it ashore to eat it.

THE WEB OF LIFE: ANIMAL PARTNERS

Many animals of different species form partnerships that help them survive. Certain fish, such as the moray eel, attract parasites that irritate their skin, gills, and mouth. That's when a cleaner fish comes in handy. The cleaner fish swims into the mouth of the eel, picks off the parasites, and eats them. Sometimes, however, the customer forgets what the cleaner fish is really there for and has it for a snack.

Another partnership is formed between the clownfish and the sea anemone. Sea anemones (ah-NEH-moh-nees) are soft-bodied animals that attach themselves to rocks and reefs. These colorful creatures look like underwater flowers. However, they have long, poisonous tentacles which they use to kill their prey. The clownfish cautiously approaches and then rubs itself against the anemone. For some reason, this prevents the anemone from stinging the clownfish, which settles in among the tentacles and waits for dinner to swim by. The clownfish feeds on prey that escapes the anemone and draws other fish into the anemone's grasp so it, too, can eat.

ENDANGERED SPECIES

Because sea turtles lay their eggs on beaches, the eggs are easily hunted and/or destroyed. Turtle eggs are a popular food among humans in many parts of the world, as are the turtles themselves. Turtles must come to the surface to breathe air, and many get caught in fishing nets which pull them below the water and they drown. Efforts are being made to save sea turtles, and over 70 conservation laws have been passed to protect them.

Many fish used as food have been caught in such large numbers that they are disappearing. Since the 1970s, the numbers of bluefin tuna have decreased by 90 percent and Atlantic swordfish have decreased by 50 percent.

In certain areas, other animals are threatened. They include the starlet sea anemone, the giant clam, the Olive Ridley turtle, the loggerhead turtle, the coelacanth, the dalmatian pelican, the West Indian manatee, the marine otter, the monk seal, and the polar bear.

A polar bear eating kelp. Although usually thought of as land animals, polar bear spend much of their time in the water in search of food. (Reproduced by permission of Corbis. Photograph by Kennan Ward.)

Also, about 10 percent of coral reefs have died because of pollution and other causes. It is estimated that by the year 2020, another 30 percent will be lost. As a result, all of the animals that depend upon the reef for food and shelter will be affected.

HUMAN LIFE

Because the water is fairly shallow and reachable by divers, the waters over the continental shelves have been explored by humans more than any other area of the ocean. Most commercial fishing areas are located over the shelves, as are those areas from which oil and natural gas are extracted.

The best-known shelves are those off the coasts of the United States, eastern Canada, western Europe, and Japan. In these places, scientific studies are routinely conducted and the information is made available to everyone. Oil companies that have worked in other areas, such as the Persian Gulf, have obtained knowledge about them. Oceanographic organizations have studied the Red Sea, the Yellow Sea, and the shelf off the coasts of Argentina and northwestern Africa. However, some developing countries have closed their waters to foreign scientists; as a result, little is known about those shelves.

> ## THE BRAINIEST MAMMAL
>
> The mammal with the largest brain is the sperm whale, a species found in all oceans. The whale's head makes up about one-third of its body. Since an average-sized male whale is about 63 feet (19 meters) long, its head would be about 20 feet (6 meters) long.

IMPACT OF THE CONTINENTAL MARGIN ON HUMAN LIFE

Because the continental margin is that area of the ocean most easily reached by humans, it has an important effect on human life.

Food About 90 percent of the world's marine food resources comes from the waters over the continental shelves. Most of those resources consist of fish.

Different methods and equipment are used to catch fish, depending upon the type of fish desired and where they are found. Some fish live in the water column and may be caught on the surface or in midwater, while others live on the ocean floor. These two categories may be divided further into fish caught at the shoreline, above the continental shelf, or in midocean.

Sport fishing is done for recreation. The catch is taken home and may be eaten, but the family does not depend upon it for food. Subsistence fishing is done to obtain food for a family or even an entire community. Extra fish may be sold to neighboring communities. Native peoples who live on islands often do subsistence fishing.

Commercial fishing is done to earn money. When carried out by small owner-operated companies, much of the work is done by hand. The equipment is usually simple and the number of fishing boats small. Industrial, large-scale fishing usually involves modern, high-powered equipment. Some huge factory ships may be as long as 330 feet (100 meters) and are equipped with the automatic machinery needed to catch, handle, store, and process huge amounts of fish.

Energy Millions of years ago, sediments from dead animal and plant life collected on the ocean bottom. Over time, these sediments fossilized (turned to stone). More time, heat, and pressure from overlying rock worked to liquify these sediments and turn them into fossil fuels, primarily gas and oil. To obtain fossil fuels, oil companies build large rigs—platforms high above the ocean surface but anchored to the sea bed. From these platforms, drilling is done into the rocky ocean floor, releasing the gas or oil, which is then pumped to shore through pipelines. Most gas and oil deposits have been obtained from offshore rigs. More than $1 billion worth of gas and oil is pumped from the continental shelves of the United States each year.

Ocean surface waters absorb large quantities of solar energy (energy from the Sun). A process known as Ocean Thermal Energy Conversion is used to capture some of that energy for human use. Conversion plants are located in Hawaii and other tropical islands. Energy from ocean currents and waves is also being explored as a source of power. One day it may be possible to anchor a turbine (energy-producing engine) in a fast-moving current, such as the Gulf Stream, and use it to produce power.

THE SNEAKY VEGETABLE

Although seaweed is popular as a food in some countries, particularly Japan, many people in the United States associate it with those slimy green things lying draped over the rocks at the beach. But most people have eaten seaweed and never realized it. As a product called carrageenan, seaweed is added to toothpaste, ice cream, gelatin, peanut butter, marshmallows, and even some meat products. It acts as a kind of glue to help hold the other ingredients together.

West Pacific natives net fishing on a raft. This type of fishing is probably done to obtain food. (Reproduced by permission of Corbis. Photograph by Amos Nachoum.)

The energy in the tides has already been harnessed in different ways. The tides are channeled into salt ponds, where salt from the water is collected for sale. The tides also lift ships in and out of drydock where repairs are made. The first tidal power station was developed in an estuary in France where turbines were built into a dam that spans the estuary. As the tides flow in and out of the estuary, they turn the turbine blades, which produce electricity.

Minerals and metals Minerals and metals are other important oceanic resources. Rocks, sand, and gravel dredged from the sea bottom, especially in the North Sea and the Sea of Japan, are used in the construction of roads and buildings. Along the Namibian coast of southwest Africa, diamonds are mined from the sea floor. Some minerals, such as sulfur, are pumped from the ocean beds as liquids.

IMPACT OF HUMAN LIFE ON THE CONTINENTAL MARGIN

The area of the ocean most affected by human action is over the continental margin.

Use of plants and animals After World War II (1939–45), the technology of commercial fishing improved and a growing population increased the demand for fish as a food source. By the 1970s, however, major food species, such as herring and cod, had been greatly reduced. As of 1995, 22 percent of marine fishing areas had been overused or depleted, and, in 44 percent of fishing areas, the maximum numbers of fish allowed by regulations were already being taken. Because most commercial fishing is done with large nets, other unwanted creatures are also caught, many of which die. For every 1 pound (2.2 kilograms) of shrimp caught in the ocean, 5 pounds (11 kilograms) of other species are thrown away.

WELLS IN THE SEA

The first offshore oil well was drilled off the coast of California in 1896. This well, and those that followed, was drilled from a pier extending from shore. The first free-standing drilling structures were not built until the 1940s.

Since that time, the design of drilling "rigs" has changed. To stand in deep water and withstand storms at sea, they require long legs and a sturdy platform. One of the tallest platforms currently in operation is off the coast of Santa Barbara, California. The distance from its "feet" anchored in the ocean floor to the top of a derrick on its platform high above the water is 1,165 feet (355 meters). This makes the rig only about 100 feet (30 meters) shorter than the Empire State Building in New York.

Exploration for wells is often done by mobile rigs positioned on drilling ships. The ships maintain their positions over the well using special propellers, and drilling in deeper waters is possible. During the 1980s, test wells were drilled from these ships in waters more than 6,500 feet (2,000 meters) deep.

Much has been said about how human greed and carelessness have threatened many species of fish. However, not all scientists agree that the problems are serious or that they lack solutions. For example, between 1983 and 1993, the total number of fish available actually increased slightly. Some scientists believe that by monitoring the numbers of fish in certain species so that they are not overfished, and by extending human consumption to more types of fish, the fish used for food need not be threatened.

Fish farms are another means of helping maintain certain species of commercially popular fish. Also called aquaculture, fish farming involves raising fish species under the best growing conditions in farms built along waterways. The fish most commonly farmed are shellfish, such as oysters, mussels, scallops, and clams. Crabs, lobsters, shrimp, salmon, trout, and tilapia are also farm raised, but to a lesser extent. The output of fish farms has tripled since 1984, and it is estimated that by the year 2000, the numbers of fish produced by fish farms will, in some cases, exceed those caught in the wild.

Other sea plants and animals are endangered because they have been collected as souvenirs or art objects. When sea shells left behind by dead animals are taken, no harm is usually done. However, many shells available commercially are taken from living animals and, as a result, the animals are left to die.

In response to these problems, marine parks and reserves have been set up all over the world to protect endangered species. They include the Shiprock Aquatic Reserve in Australia and the Hervey Bay Marine Park in California.

Natural resources Large quantities of natural resources, such as oil and minerals, can be found in the ocean water or beneath the ocean floor. These resources have not been used up because they are still too difficult or too expensive to obtain. As methods improve, however, that may change. The first areas likely to be depleted are those around the continental margins. Already, some sections of the sea floor have slumped because underlying oil and gas have been removed.

The Law of the Sea After World War II, many countries began to expand their use of the oceans. Some countries started using ocean areas other countries had already claimed for themselves, and arguments resulted. In 1994, the United Nations approved a treaty among all nations called the Law of the Sea.

The Law of the Sea Treaty maintains that a country's territory extends about 14 miles (22.2 kilometers) from its coast. That country, therefore, has a right to defend that territory. The same country also has the right to control the natural resources on its continental shelf, or for an area extending about 230 miles (370 kilometers) from its coast. This includes resources obtained from fishing and drilling for oil and natural gas. The rest of the ocean is international territory, and all countries can share in its resources.

Quality of the environment In 1996, 14,000,000,000 pounds (31,000,000,000 kilograms) of waste were dumped into the oceans. Most oceanic pollution caused by humans is concentrated along the continental margins. Sewage and industrial wastes are contributed from coastal cities, adding dangerous metals and chemicals to the water. Discarded items, such as plastic bags and old fishing nets, pose a hazard for animals that get caught in them. Medical wastes, such as needles and tubing, are also a danger.

Insecticides (insect poisons) and herbicides (weed poisons) reach the oceans when the rain washes them from fields into rivers that carry them to the sea. These poisons often enter the food chain, become concentrated in the bodies of some fish and other organisms, and are consumed by humans who eat the fish. Fertilizers and human sewage are also a problem. They cause phytoplankton to reproduce rapidly. When the plants die, their decaying bodies feed bacteria. The bacteria rapidly multiply and use up the oxygen in the water, and other organisms, such as fish, soon die.

Oil spills from tanker ships and leaks from pipelines and offshore oil wells are also major pollutants to oceans. Power plants and some industries often dump warm water into the oceans, causing thermal (heat) pollution. Organisms that require cooler water are killed by the heat.

A worker trying to absorb the oil from a California beach after an oil spill. (Reproduced by permission of Corbis-Bettmann.)

Agriculture, construction, and removal of trees on land dig up the soil. Often the rain then washes this loose soil into streams and rivers. Eventually, it enters the oceans and collects as sediment in coastal areas. This kills some organisms, such as clams, that cannot survive in heavy sediment.

Not all the materials that humans put into the oceans are harmful, however. Many organic materials, such as body wastes and fertilizers, actually add nutrients to the waters. Although some species of animals may not benefit from their effects, others do.

THE FOOD WEB

The transfer of energy from organism to organism forms a series called a food chain. All the possible feeding relationships that exist in a biome make up its food web. In the ocean, as elsewhere, the food web consists of producers, consumers, and decomposers. These types of organisms all transfer energy within the ocean environment.

Phytoplankton are the primary producers in the oceans. They produce organic materials from inorganic chemicals and outside sources of energy, primarily the Sun. Producers are sometimes called autotrophs, meaning "self-feeders." Green plants are an example of producers because they manufacture the compounds they need through photosynthesis.

Zooplankton and other animals are consumers. Zooplankton that eat only plants are primary consumers in the oceanic food web. Secondary consumers eat the plant-eaters and include zooplankton that eat other zooplankton. Tertiary consumers are the predators, such as tunas and sharks. Humans fall into this category. Humans are also omnivores, which means they eat both plants and animals.

Decomposers feed on dead organic matter. These organisms convert dead organisms to simpler substances. Decomposers include lobsters and large petrels, as well as microscopic bacteria.

Dangerous to the oceanic food web is the concentration of pollutants and harmful organisms which become trapped in sediments where life forms feed. These life forms are fed upon by other life forms, and at each step in the food chain the pollutant becomes more concentrated. Finally, when humans eat these sea animals, they are in danger of serious illness. Diseases such as cholera, hepatitis, and typhoid, can also survive and accumulate in certain sea animals. These diseases can then be caught by humans who eat the infected animals.

SOURCES OF OCEANIC OIL POLLUTION

Source	Percentage of Total
Waste from metropolitan (city) areas and industries	36.8
Ocean-going transportation	45.3
Offshore oil drilling	1.6
Pollutants in the air	9.4
Natural sources	7.8

QUEENSLAND, AUSTRALIA AND THE GREAT BARRIER REEF

The Great Barrier Reef, the longest structure in the world created by living organisms, consists of more than 2,500 smaller reefs joined together. The area it covers is approximately as large as the state of Kansas. At least 500,000 years old, the reef can be seen from space and was first mapped by the *Apollo 7* astronauts in 1968.

Located at the edge of Australia's continental shelf, the Great Barrier Reef stands in water from 325 to 650 feet (100 to 200 meters) deep. On the landward side is a lagoon. On the seaward side, the continental slope plunges thousands of feet into the deep-ocean basin.

> ## QUEENSLAND, AUSTRALIA AND THE GREAT BARRIER REEF
>
> **Location:** The Coral Sea off the northeast coast of Australia
>
> **Area:** 80,000 square miles (207,200 square kilometers)

Most of the algae found on the reef live within the bodies of the corals, often leaving the corals to float independently in the water while absorbing sunlight. Red algae, green algae, and kelp grow among the coral skeletons. Grasses and shrubs can be found on many cays, as well as banyan and breadfruit trees. The tree most commonly found is the mangrove.

The reef is home to millions of living creatures, including corals, sea urchins, sea slugs, oysters, and clams. The largest type of clam in the world is found on the Great Barrier Reef. It weighs up to 500 pounds (227 kilograms) and has a diameter of up to 4 feet (1.2 meters). Many species of sea turtles come to the reef to lay their eggs. Parrot fish, squirrelfish, trumpet fish, lionfish, coral trout, and moray eels also make the reef their home.

Seabirds nest on the cays and islands. More than 100,000 terns flock together annually on Raine Island. Frigate birds, gannets, and sea eagles can be seen skimming over the waters in search of food.

The sands that line the beaches are valuable to industry, and some people believe that oil lies beneath the shelf. To prevent the reef's destruction, the Australian government established the Great Barrier Marine Park in 1975. The park is the world's largest protected marine area. Visitors are controlled, many areas are reserved only for study, and bird and turtle breeding areas found in the park are closed during breeding season so people don't disturb the animals.

NORTHWESTERN AUSTRALIA AND SOUTHEAST ASIA

On the continental shelf off the coast of northwestern Australia are the islands of Sumatra, Java, Borneo, and the Malaysian Peninsula, the most volcanic islands in the world. Other islands, including New Guinea, New Zealand, and the Philippines, are formed from part of the shelf. South of Java, the slope suddenly plunges into the Indonesian Trench, a deep, steep-sided valley in the ocean floor.

This continental shelf holds the most extensive beds of seagrass in the world, including such species as wireweed and paddleweed. Anchored by the grass, sandbanks over 33 feet (10 meters) thick and many miles (kilometers) long have built up over time.

> **NORTHWESTERN AUSTRALIA AND SOUTHEAST ASIA**
> **Location:** Indian and Pacific Oceans
> **Area:** Approximately 140,000 square miles (300,000 square kilometers)

Blue swimmer-crabs, pen shells, fan mussels, green turtles, and sea snakes are just a few of the marine animals that live over the shelf. Coastal waters also yield herring, salmon, sardines, snapper, swordfish, and tuna.

A large colony of dugongs, plant-eating relatives of the manatee, are attracted to the ample supply of seagrass and make this area their home. The dugong population is declining, however, because they are hunted by native peoples for food and are often killed by boat propellers.

In the shallow waters over the shelf, petroleum and natural gas are being extracted. Pearls are also harvested here, although in sharply declining volume.

In 1986 the member nations of the South Pacific Forum declared the area a nuclear-free zone in an attempt to halt nuclear testing and prevent the dumping of nuclear waste.

WESTERN UNITED STATES

The continental margin off the northwestern coast of the United States was partly formed by a dam of rock thrust up by earthquake action about 25,000,000 years ago. Where the dam breaks the surface of the water, it forms the Farallon Islands off San Francisco. The shelf is so narrow here—only about 1 mile (1.6 kilometers) wide—that the heads of many submarine canyons reach almost to the shoreline. Large quantities of sand from the beaches are carried by currents down the canyon walls.

> **WESTERN UNITED STATES**
> **Location:** Pacific Ocean
> **Area:** Approximately 173,700 square miles (450,000 square kilometers)

Rainfall and runoff from rivers occur only during the winter months. The California Current travels south in the summer, and in winter the Davidson Current appears and moves north.

The North Pacific waters are rich in marine life, including sponges, flying fish, sharks, manta rays, seals, and many species of whales.

Commercial fishing is done all along the coast but primarily in the waters around Alaska. Oil and gas are also obtained from the area.

NORTHEASTERN UNITED STATES

Between 270,000,000 and 60,000,000 years ago, a large dam created by earthquake action was formed off the northeastern coast of the United States. A trench on its landward side was gradually filled in with sediment over time, and this area is now part of the continental shelf. Sediment that spilled over the top formed a new continental slope that is unstable and subject to landslides. Few submarine canyons are found here, although the Hudson Canyon, which is associated with the Hudson River, extends from the river's mouth into the ocean.

> ### NORTHEASTERN UNITED STATES
> **Location:** Atlantic Ocean
> **Area:** 71,410 square miles (185,000 square kilometers)

Changes in sea level have altered the appearance of the coastline and shelf. Fifteen thousand years ago, sea level was much lower, and much of the continental shelf in this area was exposed. Gradually, as sea levels rose, the shelf was covered by water.

Numerous rivers empty into the Atlantic here along this shelf, and the presence of fresh water reduces the ocean's salinity. The warm Gulf Stream current affects water temperature and circulation.

Waters here support a variety of marine life, including sea slugs, starfish, mussels, many species of crabs and lobsters, poisonous toadfish, and sperm whales.

Commercially important fish found along this shelf include ocean perch, cod, and haddock. Oil and gas are obtained from the shelf off the coast of Newfoundland and farther north.

THE NORTH SEA AND WESTERN EUROPE

The continental shelf off the coast of Europe holds the entire United Kingdom (England and its islands). Waters over this shelf include the North Sea, the Skagerrak (an arm of the North Sea), and the English Channel. Under the North Sea the shelf slopes gently downward from south to north. Although the southernmost areas are shallow, a deep canyon lies off the mouth of the Humber River, an estuary on the east coast of England. In the northeast is the Norway Deep, or Trough, an underwater canyon running parallel to the Norwegian coast and into the Skagerrak. Other troughs lie to the west of Ireland. In the north, the coastline is marked by steep cliffs.

> ### THE NORTH SEA AND WESTERN EUROPE
> **Location:** The North Atlantic
> **Area:** 212,300 square miles (550,000 square kilometers)

Tides are important in the North Sea because they influence ocean traffic. Along the coast of Norway, the tidal range (difference between high tide and low tide) is often less than 3 feet (1 meter). On the French side of the English Channel, tidal ranges of more than 27 feet (8 meters) are common. Tidal currents deposit sand in areas along the coast, causing problems in shipping routes. Several large rivers, such as the Rhine and the Elbe, lower the salinity of the water and create currents where they flow into the sea.

Sediments from the rivers and upwelling from deep, cold waters makes the North Sea rich in nutrients. The rocky shelf supports much plant life, such as algae, kelp, and eel grass. Invertebrates like cockles, mussels, scallops, sponges, and snails also thrive in the rocky areas.

More than 170 species of fish live in the North Sea, including sharks, rays, herring, mackerel, haddock, cod, and sole. The fish attract large numbers of sea birds, including puffins, gannets, terns, gulls, and ducks. Among mammals living here are gray seals, harbor seals, dolphins, porpoises, and killer whales.

Fishing has been done in the area since A.D. 500. Herring, haddock, plaice, cod, and whiting are commercially important. Atlantic salmon are farmed in some areas.

The North Sea is also the site of offshore oil and gas drilling. By 1989, 149 oil platforms were operated in the North Sea by British, Dutch, Norwegian, Danish, and German companies.

More sand and gravel is removed from the North Sea—24.5 million tons (22 metric tons) annually—than anywhere else in the world. These operations occur close to shore in waters less than 115 feet (35 meters) deep. Lime, another mineral, is mined from the seabed (sea floor). All of these materials are used as construction materials.

THE PERSIAN GULF

The Persian Gulf lies between the Arabian Peninsula and Iran. Millions of years ago, it was much larger, but the floor beneath the Persian Gulf is shrinking as the floor beneath its neighbor, the Red Sea, on the opposite side of the peninsula, expands. It is estimated that after another 50,000 years, the Persian Gulf will be completely closed as the peninsula is pushed toward Iran. At one time the floor of the gulf was above sea level. Now it is part of the continental shelf. No earthquakes or volcanoes are found here, and the seafloor has both muddy and sandy areas.

Surrounded by desert, the water in the Persian Gulf is warm, salty, and only about 330 feet

> ### THE PERSIAN GULF
> **Location:** The Arabian Sea, between Saudi Arabia and Iran
> **Area:** 92,640 square miles (240,000 square kilometers)

(100 meters) deep. Although a small amount of fresh water flows in from the Tigris and Euphrates Rivers, the climate is so hot that more water is lost from evaporation than is gained.

Large stands of mangrove trees and beds of seagrass are found in the gulf. The mud in which they grow is low in oxygen, so the thickets are not as well developed as those found in other areas.

Few coral reefs are found here, because the water is too warm. However, the gulf is home to many kinds of shellfish, such as mussels, shrimp, and oysters. Fish like sardines, anchovies, mackerel, and barracuda are also found here. The largest fish living in the area is the whale shark, which may reach 40 feet (12 meters) in length. Common birds of the gulf include terns, ospreys, and the fish-eating eagle, which travels several thousand miles to Scotland to nest. Porpoises are common, and the narwhal is seen occasionally.

The Persian Gulf has been important for oil production since 1935, and oil platforms dot its surface. Salt, as well as other minerals and metals, such as copper and zinc, are also found here and are important commercially.

Sea cucumbers, shellfish, sardines, and anchovies are all important to the fishing industry. Bahrain, a tiny island in the gulf, has been the source of pearl oysters for 2,000 years, although the pearl industry has declined since it became possible to artificially stimulate pearl growth.

HUDSON BAY

The continental shelf underlying Hudson Bay extends from the Canadian Provinces of Quebec, Manitoba, Ontario, and the Northwest Territories. Its depth ranges from 120 to 600 feet (36 to 182 meters). A deep canyon cuts through the bay and extends toward Hudson Strait.

> **HUDSON BAY**
> **Location:** Northeastern Canada
> **Area:** 480,000 square miles (768,000) square kilometers

The cold waters in Hudson Bay originate in the Arctic. Because many rivers feed into the bay, the waters are low in salt. From January until May, the bay is covered with floating ice, which in northern regions, is slow to melt and makes navigation difficult.

Many fish live in the bay, including cod and salmon. Seabirds that are found here include ducks, geese, loons, and ptarmigans. Whales also frequent the bay, and at one time were hunted here. Native peoples still hunt and fish in the area.

During the last half of the twentieth century, the bay and strait became a popular shipping route for goods going to England. Oil and natural gas are being pumped from beneath the shelf in the northern regions, making the shelf important commercially.

FOR MORE INFORMATION

BOOKS

Baines, John D. *Protecting the Oceans.* Conserving Our World. Milwaukee, WI: Raintree Steck-Vaughn, 1990.

Carwardine, Mark. *Whales, Dolphins, and Porpoises.* See and Explore Library. New York: Dorling Kindersley, 1992.

Lambert, David. *Seas and Oceans.* New View. Milwaukee, WI: Raintree Steck-Vaughn Company, 1994.

Markle, Sandra. *Pioneering Ocean Depths.* New York: Atheneum, 1995.

Siy, Alexandra. *The Great Astrolabe Reef.* New York: Dillon Press, Macmillan Publishing Company, 1992.

ORGANIZATIONS

American Cetacean Society
PO Box 1391
San Pedro, CA 90731
Phone: 310-548-6279; Fax: 310-548-6950
Internet: http://www.acsonline.org

American Littoral Society
Sandy Hook
Highlands, NJ 07732
Phone: 732-291-0055

American Oceans Campaign
725 Arizona Avenue, Suite. 102
Santa Monica, CA 90401
Phone: 800-8-OCEAN-0
Internet: http://www.americanoceans.org

Astrolabe, Inc.
4812 V. Street, NW
Washington, DC 20007

Center for Marine Conservation
1725 De Sales Street, NW
Washington, DC 20036
Phone: 202-429-5609; Fax: 202-872-0619

Environmental Defense Fund
257 Park Ave. South
New York, NY 10010
Phone: 800-684-3322; Fax: 212-505-2375
Internet: http://www.edf.org

Environmental Network
4618 Henry Street
Pittsburgh, PA 15213
Internet: www.environlink.org

Environmental Protection Agency
401 M Street, SW
Washington, DC 20460
Phone: 202-260-2090
Internet: http://www.epa.gov

Friends of the Earth
1025 Vermont Ave. NW, Ste. 300
Washington, DC 20003
Phone: 202-783-7400; Fax: 202-783-0444

Greenpeace USA
1436 U Street NW
Washington, DC 20009
Phone: 202-462-1177; Fax: 202-462-4507
Internet: http://www.greenpeaceusa.org

Project Reefkeeper
1635 W Dixie Highway, Suite 1121
Miami, FL 33160

Sierra Club
85 2nd Street, 2nd fl.
San Francisco, CA 94105
Phone: 415-977-5500; Fax: 415-977-5799
Internet: http://www.sierraclub.org

World Meteorological Organization
PO Box 2300
41 Avenue Guiseppe-Motta
1211 Geneva 2, Switzerland
Phone: 41 22 7308411; Fax: 41 22 7342326
Internet: http://www.wmo.ch

World Wildlife Fund
1250 24th Street NW
Washington, DC 20037
Phone: 202-293-4800; Fax: 202-293-9211
Internet: http://www.wwf.org

WEBSITES

Note: Website addresses are frequently subject to change.

Discover Magazine: http://www.discover.com

Journey North Project: http://www.learner.org/jnorth

Monterey Bay Aquarium: http://www.mbayaq.org

National Center for Atmospheric Research: http://www.dir.ucar.edu

National Geographic Society: http://www.nationalgeographic.com

National Oceanic and Atmospheric Administration: http://www.noaa.gov

Scientific American Magazine: http://www.scientificamerican.com

Time Magazine: http://time.com/heroes

BIBLIOGRAPHY

Cerullo, Mary M. *Coral Reef: A City That Never Sleeps.* New York: Cobblehill Books, 1996.

Grolier Multimedia Encyclopedia. Danbury, CT: Grolier, Inc., 1995.

Engel, Leonard. *The Sea.* Life Nature Library. New York: Time-Life Books, 1969.

Gutnick, Martin J., and Natalie Browne-Gutnik. *Great Barrier Reef.* Wonders of the World. Austin, TX: Raintree Steck-Vaughn Publishers, 1995.

Holing, Dwight. *Coral Reefs.* Parsippany, NJ: Silver Burdett Press, 1995.

Macquitty, Miranda. *Ocean.* Eyewitness Books. New York: Alfred A. Knopf, 1995.

McLeish, Ewan. *Oceans and Seas.* Habitats. Austin, TX: Raintree Steck-Vaughn Company, 1997.

Morgan, Nina. *The North Sea and the Baltic Sea.* Seas and Oceans. Austin, TX: Raintree Steck-Vaughn Company, 1997.

The Ocean. Scientific American, Inc., San Francisco: W. H. Freeman and Company, 1969.

"Ocean." *Encyclopaedia Britannica.* Chicago: Encyclopaedia Britannica, Inc., 1993.

Oceans. The Illustrated Library of the Earth. Emmaus, PA: Rodale Press, Inc., 1993.

Ocean World of Jacques Cousteau. Guide to the Sea. New York: World Publishing, 1974.

Pernetta, John. *Atlas of the Oceans.* New York: Rand McNally, 1994.

Ricciuti, Edward R. *Ocean.* Biomes of the World. Tarrytown, NY: Benchmark Books, 1996.

Rosenblatt, Roger. "Call of the Sea." *Time* (October 5, 1998): 58-71.

Wells, Susan. *The Illustrated World of Oceans.* New York: Simon and Schuster, 1991.

Whitfield, Philip, ed. *Atlas of Earth Mysteries.* Chicago: Rand McNally, 1990.

A coniferous forest in Paradise Bay, Antarctica. [©1995 PhotoDisc.]

A Clark's Nutcracker perched atop a pine tree.
[©1995 PhotoDisc.]

Parchment fungi growing on the ground of a coniferous forest. [©1995 PhotoDisc.]

A red fox prancing through a snow-covered coniferous forest. [©1995 PhotoDisc.]

Snakes are just one of the many species of reptiles commonly found in coniferous forests. [©1995 PhotoDisc.]

In autumn, tree leaves in deciduous forests provide a scene of brilliant colors. [Reproduced by permission of Field Mark Publications.]

Tree frogs like this one are commonly found among deciduous forest trees. [©1995 PhotoDisc.]

A caterpillar feeds on the leaf of a deciduous tree. [Reproduced by permission of Field Mark Publications.]

A single mushroom found on a deciduous forest floor. [©1995 PhotoDisc.]

Clover and ferns are just some of the vegetation found in deciduous forests. [©1995 PhotoDisc.]

A great horned owl resting on a tree branch. [©1995 PhotoDisc.]

Monument Valley found in the deserts of Arizona and Utah. [Reproduced by permission of Archive Photos, Inc.]

The massive landform of Ayers Rock found in the Australian desert. [Reproduced by permission of Archive Photos, Inc.]

Vultures pick at the flesh of a dead giraffe in an African desert. [Reproduced by permission of JLM Visuals.]

A sunset illuminates a cactus in one of the world's deserts. [©1995 PhotoDisc.]

A closeup of the blooms on a cactus, one of the major plant species in a desert. [©1995 PhotoDisc.]

The hyena, commonly found in the desert, has one of the strongest set of jaws of any animal. [©1995 PhotoDisc.]

These waves, called breakers, collapse on a shoreline because the water at the bottom of them are slowed by friction as they roll along the shore. The top of the waves then outrun the bottom and topple over in a heap of bubbling foam. [©1995 PhotoDisc.]

Seagulls flying high above the ocean. They often eat fish and other small creatures found in the ocean's continental margin. [©1995 PhotoDisc.]

A close-up of seagrass and kelp that grow in the continental margins and provide nutrition for many of the animals that live there. [Reproduced by permission of Field Mark Publications.]

Although sea turtles live in the continental margin, they lay their eggs in the sand along the shore. [©1995 PhotoDisc.]

Although they resemble fish, dolphins are one of very few mammals found in the continental margin. [©1995 PhotoDisc.]

Octopi, like this one, are commonly found among the rocks and sand of the continental margin. [Reproduced by permission of JLM Visuals.]

DECIDUOUS FOREST

A tree is a large woody plant with one main stem, or trunk, and many branches that lives year after year. A forest is a large number of trees covering not less than 25 percent of the area where the tops of the trees, called crowns, interlock forming an enclosure or canopy when the trees mature. This chapter is about deciduous (dee-SID-joo-uhs) forests. Deciduous trees, such as oaks, basswoods, and elms, are those that lose their leaves during cold or very dry seasons, as compared to evergreen trees that keep their leaves year-round, usually for several years at a time. For information about evergreen forests, see the chapters titled "Coniferous Forest" and "Rain Forest."

Temperate deciduous forests grow in areas with cold winters and warm summers. They are found primarily in the Great Lakes region and the eastern half of the United States, parts of central and western Europe, parts of Russia, and parts of Japan and China. Tropical deciduous forests grow in areas around the equator where the weather is warm. There are two types of tropical deciduous forests—those that grow in dry climates and those that grow in moist climates. Those with dry climates occur primarily in central India; parts of Brazil; and on the African continent from Angola to Tanzania, northward to the Sudan, and over much of West Africa. Those with moist climates are found primarily in northeastern Australia, eastern India, and parts of Burma, Thailand, and Indonesia.

HOW DECIDUOUS FORESTS DEVELOP

Forests evolved during Earth's prehistoric past. Since then, all forests have developed in essentially the same way, by means of a process called succession.

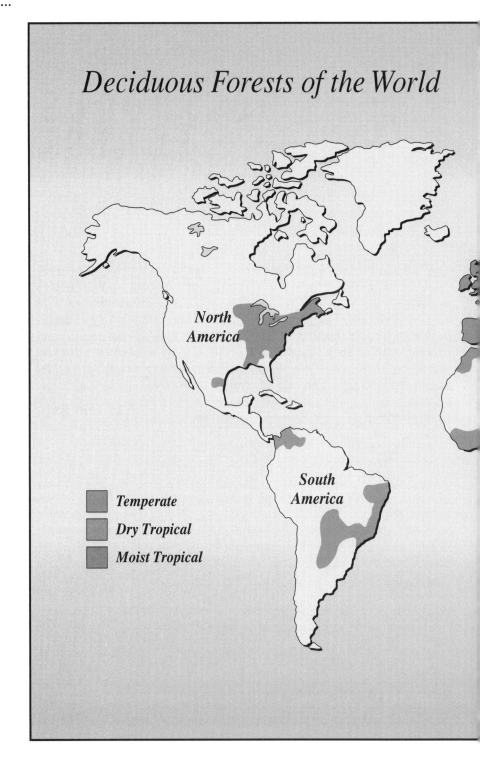

Deciduous Forests of the World

North America

South America

Temperate

Dry Tropical

Moist Tropical

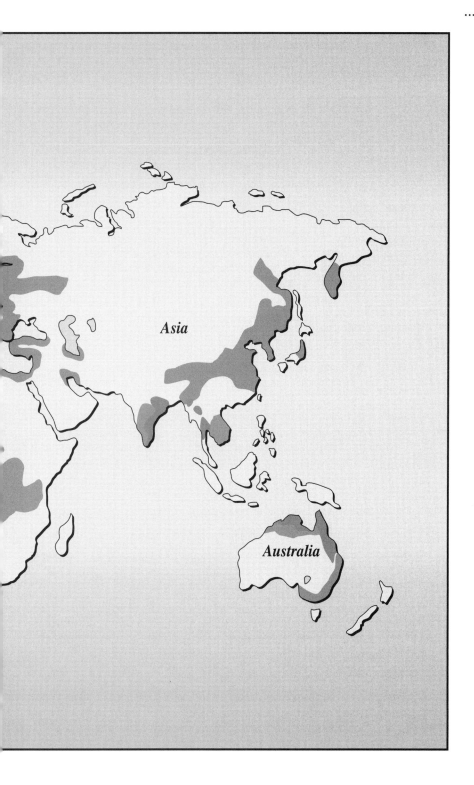

Asia

Australia

THE FIRST FORESTS

The first forests evolved from ferns and other prehistoric plants that, over time having adapted to the surrounding environment, grew more tree-like. Trees that preferred a warm, humid, tropical climate developed first, followed by those that were adapted to drier, cooler weather. The first deciduous trees evolved about 180,000,000 years ago.

About 1,000,000 years ago, during the great Ice Ages, glaciers (slow-moving masses of ice) covered much of the planet and destroyed many of the world's forests. By the time the glaciers finally retreated to where they are today, about 10,000 years ago, they had scoured the land of plants. As time passed, deciduous birch trees were among the first to return to the regions once covered by ice. In fact, the years between 8,000 and 3,000 B.C. are often referred to as the "Birch Period." Birches prepared the way for other trees.

> ### KILLED BY THE FROST
>
> During prehistoric times, before the great Ice Ages, North America and northern Europe were covered in deciduous trees such as walnut, hickory, sycamore, oak, maple, and chestnut. The climate was warmer than that of present times. As the ice advanced, however, the more delicate trees were forced to retreat to the south. In North America, this was possible, because the mountain ranges run north and south, which allowed the trees to spread along the valleys between them. In Europe, however, the mountains tend to run east and west, so the trees were trapped. For this reason, many European species did not survive.

SUCCESSION

Trees compete with one another for sunlight, water, and nutrients, and a forest is constantly changing. The process by which one type of plant or tree is gradually replaced by others is called succession. Succession took place following the last Ice Age, just as it takes place today after the land has been stripped of vegetation from causes such as forest fires. Succession produces different types of forests in different regions, but the process is essentially the same everywhere. During succession, different species of trees become dominant as time progresses and the environment changes.

Primary succession Primary succession usually begins on bare soil or sand where no plants grew before. When the right amount of sunlight, moisture, and air temperature are present, seeds begin to germinate (grow). These first plants are usually made up of the grasses and forbs (a nonwoody broad-leaved plant) type. They continue to grow and eventually form meadows. Over time, and as conditions change, other plants begin to grow such as shrubs and trees, which become dominant and replace or take over where the grasses and forbs originally grew.

As primary succession continues, "pioneer" trees—birch, pine, poplar, and aspen—begin to thrive. These are all tall, sun-loving trees, and they quickly take over the meadow. They also change the environment by making shade. Now trees with broader leaves that prefer some protection from the sun, such as red oaks, can take root. If conditions are right, a mixed forest of sun-loving and shade-loving trees may continue for many years. Eventually, however, more changes occur.

The climax forest Seedlings from pioneer trees do not grow well in shade; therefore, new pioneer trees do not grow. As the mature trees begin to die from old age, disease, and other causes, the broad-leaved trees become dominant. However, the shade from these broad-leaved trees can be too dense for their own seedlings, as well. As a result, seedlings from the trees that prefer heavy shade, such as beech and sugar maple, begin to thrive and eventually dominate the forest. These trees produce such deep shade that only those trees or plants that can survive in complete shade will succeed there. When this happens, the result is a climax forest—one in which only one species of tree is dominant. If one tree dies, another of the same species grows to take its place. In this way, a climax forest can endure for thousands of years.

Few true climax forests actually exist, however, because other changes take place that interfere with a forest's stability. Fires, floods, high winds, and people can all destroy a single tree to acres of trees. Glaciers can mow them

An illustration showing forest succession, from submergents to tree stage.

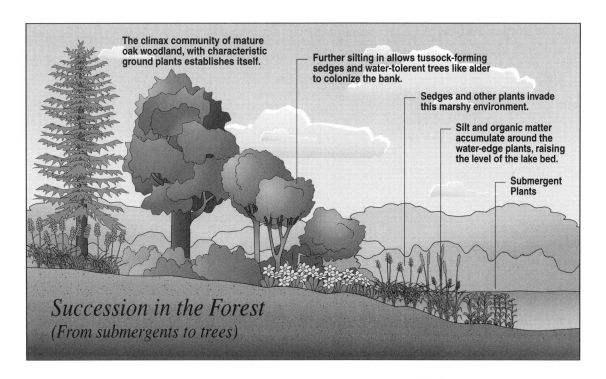

The climax community of mature oak woodland, with characteristic ground plants establishes itself.

Further silting in allows tussock-forming sedges and water-tolerant trees like alder to colonize the bank.

Sedges and other plants invade this marshy environment.

Silt and organic matter accumulate around the water-edge plants, raising the level of the lake bed.

Submergent Plants

Succession in the Forest
(From submergents to trees)

down, and volcanoes can smother them with ash or molten rock or knock them over with explosive force. Then the process of succession starts over.

Secondary succession When the land has been stripped of trees, it will eventually be covered with them again if left alone. This is called secondary succession and can take place more quickly than primary succession. Seeds from other forests in neighboring regions are blown by the wind or carried by animals to the site. Soon, the seeds take root and seedlings sprout, and the process begins again.

KINDS OF DECIDUOUS FORESTS

In general, deciduous forests are broadly classified according to the climate of the region in which they grow—temperate, dry tropical, or moist tropical.

TEMPERATE DECIDUOUS FOREST

This pine forest in Wisconsin is an example of a climax forest where only one species of tree is dominant. (Reproduced by permission of JLM Visuals.)

Temperate climates are moderate, and temperate deciduous forests can be categorized in terms of the species of trees that are most common. The

following three types are common in North America, and similar species predominate in Europe and Asia:

- Beech-maple: American beech and sugar maple forests mixed with some conifers (trees bearing cones) are found in the southern Great Lakes region, in New England, and in southeastern Canada.

- Oak-hickory: In the southern United States, oaks, such as white oak and chestnut oak, are dominant. Farther west, red oak and black oak are mixed with hickory.

- Mixed mesophytic: Mesophytic means a forest that requires only a moderate amount of water, such as the forests in the Appalachian Mountains. Mixed mesophytic forests can be dominated by any of ten or more species of trees, such as buckeye, magnolia, birch, ash, black cherry, and sugar maple.

DRY TROPICAL DECIDUOUS FOREST

Dry tropical deciduous forests occur in regions that have long, severe dry seasons, such as the savannas (grasslands) of Africa. The tallest trees are shorter and more twisted than those in a temperate forest, and the bark is thicker. Some trees may store water in their trunks. During the dry season, the trees lose their leaves, and many produce flowers. Some, such as wattle trees in Australia, have adapted by producing thorns instead of many leaves.

MOIST TROPICAL DECIDUOUS FOREST

Moist deciduous forests are found in tropical regions where both rainy and dry seasons occur during the year. These forests are different from tropical rain forests in that the trees, including teak and rosewood, are not as tall and have rather thick bark. In the first month of the dry season, the leaves fall. New leaves sprout just before the rainy season begins.

ATTACK OF THE KILLER WALNUT

Life in the forest is a constant struggle for food and living space, even among the trees. All around the foot of almost any tree are the seedlings of other trees, ready to take over the space if they can. But some trees put up a fight. The black walnut, for instance, has a powerful poison in its leaves, roots, and nut shells. This poison kills the roots of any plants trying to grow in the same soil. Even if the tree is removed, the poison remains in the soil for a long time.

CLIMATE

The three types of climates in which deciduous trees grow are temperate, dry tropical, and moist tropical.

TEMPERATE CLIMATE

Temperate climates, such as in the northeastern United States, are moderate. Winters are cold but not severe and summers are warm but with few extremely hot days. January temperatures average between 10°F (-12°C) and 30° F (-1.1°C) and July temperatures between 36°F (2.2°C) and 81°F (27°C).

There is plenty of moisture and few long, dry periods. A temperate forest usually requires at least 20 inches (51 centimeters) of precipitation (rain, snow, or sleet) each year, and most of these forests average between 30 to 80 inches (80 to 200 centimeters). Humidity is usually between 60 and 80 percent, and this moist blanket of air helps keep the temperatures from becoming extreme. However, the weather in a temperate climate is usually unpredictable. Although precipitation usually occurs throughout the year, much of it may fall as part of a severe storm system.

Temperate climates have four distinct seasons—spring, summer, fall or autumn, and winter. Warm weather begins in spring and temperatures gradually increase through the summer months. In the fall, the cooler weather takes over, and temperatures gradually drop to winter's lows. It is during the cooler months in the autumn that the leaves of deciduous trees turn glorious colors and then drop off. In the winter, the trees are bare.

Snowshoeing through a bare, snow-covered deciduous forest. The blanket of snow keeps the soil warm allowing bacteria and other organisms to continue to break down dead plant matter. (Reproduced by permission of Corbis. Photograph by Robert Holmes.)

DRY TROPICAL CLIMATE

In dry tropical climates, such as that in parts of India and Africa, the drop in precipitation is usually accompanied by hotter temperatures. The length of the dry period and the level of heat help determine the type of trees that will grow there. In Tanzania, for example, severe dry periods may last up to seven months. Precipitation is often undependable, and, when it does occur, it may fall in large amounts. In India, for example, as much as 35 inches (89 centimeters) of rain may fall in a single 24-hour period. Leaves fall from the trees at the start of the dry period and grow again when the rains come.

MOIST TROPICAL CLIMATE

In tropical regions where moist climates occur, such as Burma and parts of India, temperatures remain high, but precipitation is more regular and predictable. In a rain forest, rain falls year-round, but in a moist tropical climate, dry periods occur. However, the dry periods are not as long or severe as in a dry tropical climate. As in dry climates, the trees shed their leaves when the dry period begins. New growth begins with the rains.

GEOGRAPHY OF DECIDUOUS FORESTS

The geography of deciduous forests includes landforms, elevation, soil, and water resources.

LANDFORMS

In the Northern Hemisphere, the landscape over which temperate deciduous forests grow includes mountains, valleys, rolling hills, and flat plateaus.

In the Southern Hemisphere, dry deciduous forests tend to occur near grasslands where the land is rolling or more nearly level. Moist deciduous forests are often found on mountainsides or rolling hills.

ELEVATION

On mountainsides, deciduous trees grow up to elevations where the coniferous trees begin, at approximately 9,000 feet (2,743 meters) above sea level. They are not well adapted to the colder, drier conditions higher up. Below this, deciduous forests are found at many different elevations.

SOIL

Because it receives a new blanket of leaves every autumn, the soil in temperate deciduous forests tends to be deep and rich with a wide variety of nutrients. The foliage of low-lying plants also dies off and adds its own nutrient characteristics to the soil. In winter, snow blankets the ground and, beneath its protective layer, bacteria, earthworms, and insects continue to break down the dead vegetation, creating a dark humus. (Humus is the spongy matter produced when the remains of plants and animals are broken down. It contains chemicals, like nitrogen, that are vital to plant growth, and it is able to absorb water.) Oak leaves are difficult to break down; therefore, soil beneath oak trees is not as rich.

In tropical climates, the soil differs from region to region. Soils in dry grasslands are sandy and dusty. Because of the long dry periods between the rainy seasons, dead plant matter does not have a chance to decompose (break down) and release nutrients, making the soil less rich. Moist tropical regions may also have poor soil because any topsoil is often washed away during heavy rains. Also, the shade is so dense that few smaller plants may grow there.

In general however, the presence of trees protects soil from erosion by holding it in place. Fallen trees are important in conserving and cycling nutrients and in reducing erosion. Trees also create windbreaks, which helps prevent topsoil from being blown away.

WATER RESOURCES

In temperate regions, water resources include rivers, streams, springs, lakes, and ponds. In tropical regions, rivers and seasonal streams are often the primary sources of water.

A fungus creates a circular pattern on fallen maple leaves and speeds forest decay. This decomposition enriches the soil and enables new growth. (Reproduced by permission of Corbis. Photograph by Gary Braasch.)

PLANT LIFE

Most forests contain a mixture of several types of trees and plants, and deciduous forests are no different. Stands (groups) of coniferous and non-coniferous evergreen trees may exist within their boundaries.

The trees and smaller plants in a forest grow to different heights, forming "layers." The crowns of the tallest trees create a canopy, or roof, over the rest. In the deciduous forest, the tallest trees are often oaks and hickories. Beneath their canopy grow shorter, shade-tolerant trees, such as maples. This shorter layer is called the understory. The next layer, only a few feet off the ground, is composed of small shrubs, such as junipers, and some flowering plants. The very lowest layer consists of small plants that live atop the soil.

Plant life within most deciduous forests includes not only trees but also bacteria; algae, fungi, and lichens; and green plants other than trees.

ALGAE, FUNGI, AND LICHENS

Algae (AL-jee), fungi (FUHN-jee), and lichens (LY-kens) do not fit neatly into either the plant or animal category. In this chapter, however, they will be discussed as if they, too, were plants.

Algae Most algae are single-celled organisms, although a few are multicellular. Certain types of algae have the ability to make their own food. During a process called photosynthesis (foh-toh-SIHN-thuh-sihs), they use the energy from sunlight to change water and carbon dioxide (from the air) into the sugars and starches they require as food. Other algae may absorb nutrients from their surroundings. Although most algae are water plants, blue-green algae do appear in woodlands. They survive as spores (single cells that have the ability to grow into a new organism) during dry periods and return to life as soon as it rains.

Fungi Fungi cannot make their own food by means of photosynthesis. Some, like mold and mushrooms, obtain nutrients from dead or decaying organic (material derived from living organisms) matter. They assist in decomposition of this matter, thereby releasing the nutrients needed by other plants. Other types of fungi are parasites and attach themselves to other living things. Fungi also reproduce by means of spores.

Fungi prefer a moist, dim environment, and they thrive in shadowy forests of temperate regions. Common types of fungi include jack-o'-lantern, pear-shaped puffball, fawn mushroom, turkey tail, and destroying angel. Some, like chantarelles, grow directly on the soil, while others grow on the trunks of trees. Many, such as the fly agaric, are mushroom-like. Others, such as tree ears, resemble a porch roof as they protrude from tree trunks. One type, the mycorrhiza, surround the roots of certain trees, such as beeches and oaks, and help the roots absorb nutrients from the soil.

Lichens Lichens are actually combinations of algae and fungi that live in cooperation. The fungi surround the algae cells. The algae obtain food for themselves and the fungi by means of photosynthesis. It is not known if the fungi aid the algal organisms, although they appear to provide them with protection and moisture.

Lichens often appear on rocks and other bare woodland surfaces. They are common in all types of forests and seem able to survive most climatic conditions. Lichens, however, will not grow in the presence of air pollution. For this reason, they are a good indicator of air quality.

GREEN PLANTS OTHER THAN TREES

Most green plants need several basic things to grow: sunlight, air, water, warmth, and nutrients. In deciduous woodlands, water and warmth are often abundant, at least seasonally. Light may be more scarce, although deciduous trees tend to be well-spaced, allowing sunlight to reach the forest floor. The remaining nutrients—primarily nitrogen, phosphorus, and potassium—are obtained from the soil and may not always be in large supply.

Woodlands are home to both annual and perennial plants. Annuals live only one year or one growing season. Perennials live at least two years or two growing seasons, often appearing to die when the climate becomes too cold or too dry, but returning to "life" when conditions improve.

Growing season In temperate climates, green plants that grow close to the forest floor, such as the anemone, skunk cabbage, and trillium, appear in early spring when full sun can reach them. The primrose and the bluebell quickly bloom as the trees' leafy canopy begins to form overhead. Later, when the trees are in full leaf, only the shade-loving plants, such as fern, moss, and ivy, will thrive. Growth continues throughout the summer then stops in autumn. Annuals die, leaving their seeds to carry through the winter. Although the foliage (above ground growth such as leaves and stems) of perennials dies, their roots remain alive to send up shoots again the following spring.

In tropical climates, both dry and moist, the growing season begins just before the onset of the rainy season. It ends as the dry season begins. As in temperate climates, annuals die. Perennials usually have some method of storing water, such as a large taproot (a large center root that grows downward), that enables them to survive long periods without rain.

A NOSE FOR TRUFFLES

You may have heard of Babe, the pig that thought he was a sheepdog. But it may surprise you to learn that in France, pigs, as well as dogs and even goats, are trained to hunt truffles. A truffle is a type of edible fungi that grows under the soil in deciduous woodlands. The pigs are trained to recognize the smell of a truffle and sound the alarm when they find one. The truffles are then dug up and sold to people who like to eat the fungi.

Reproduction Most green plants reproduce by means of pollination (the transfer of pollen from the male reproductive flower part to the female). Pollen is carried by visiting animals, such as birds or insects, or by the wind which blows it from one flower to another. As the growing season comes to an end, most green plants produce seeds. The seed's hard outer covering protects it during cold winters or long dry seasons.

A few woodland plants, such as ferns, reproduce by means of rhizomes, long, rootlike stems that spread out below ground. These stems develop their own root systems and send up sprouts that develop into new plants. They also reproduce by spores which develop on the undersides of the leaf. When these spores mature, they are released from the plant and fall to settle on the soil where they begin to grow into a new fern plant.

Common deciduous forest green plants In temperate climates, many species of small shrubs, mosses, ferns, herbs, and brambles thrive beneath the trees, along the edges of the forest, and in clearings. They include wildflowers such as lady's slipper and larkspur, and shrubs such as witch hazel, sumac, and spicebush. Some shade-loving plants also grow beneath the trees. These include mayapple, jack-in-the-pulpit, poison ivy, and many species of ferns.

In moist tropical regions, a layer of smaller evergreen trees often grows beneath the deciduous canopy. Both high- and low-climbing vines, as well as most smaller plants, may in some cases be almost completely absent. In dry tropical regions, the lower layers may consist of thorny shrubs, cacti, grasses, and small palms.

A jack-in-the-pulpit in an Ontario, Canada, forest. (Reproduced by permission of Field Mark Publications. Photograph by Robert J. Huffman.)

DECIDUOUS TREES

Deciduous trees lose their leaves, which tend to be broad and flat, during cold or very dry periods. However, when temperatures are warm year-round and rainfall is constant, such as in a rain forest, these same trees may become evergreen and keep their leaves year-round. Trees are hardy perennials, and many deciduous trees live from 100 to 250 years.

The fact that most trees have a single strong stem, or trunk, gives them an advantage over smaller woody plants in that most of their growth is directed upward. Although conifers devote their energy to growing ever taller, deciduous trees spread out their limbs and branches from their trunks to create a crown of leaves.

During autumn, trees in temperate regions lose the green color in their leaves because chlorophyll, a green substance in the leaves, breaks

down. As a result, other colors present in the leaves are then visible, creating the brilliant hues of the autumn forest. This color change is brought on by shorter days, where hours of sunlight are decreased, and by cooler temperatures, which occur during autumn.

Each year, as a deciduous tree grows, its trunk is thickened with a new layer, or ring, of vessels that carry water and nutrients from the roots to the branches. When a tree is cut down, its age can be determined by how many of these rings are present. As a tree ages, the vessels from the center outward become hardened to produce a sturdy core. In temperate climates, deciduous trees may grow from 3 to 5 feet (1 to 1.5 meters) in diameter and 80 to 100 feet (25 to 30 meters) in height during their life span.

Growing season Moisture and temperature conditions are two important environmental factors affecting the growing season of deciduous trees. In temperate climates, the growing season takes place during the spring and summer and may last from five to seven months. Some trees, such as beech and maple, produce their leaves earlier in the season, while oaks produce leaves later. In the autumn, cooler temperatures make it harder to absorb the water needed to maintain broad leaves. Broad leaves also tend to lose a lot of water into the atmosphere. For these reasons, as well as other factors that include temperature and day length, the leaves die and fall.

In tropical climates, growth slows or stops during hot, dry periods and the trees drop their leaves. Growth begins again when the rain comes. When dry periods are very long, trees in these areas may be stunted and develop very thick bark, which protects against moisture loss.

Reproduction In general, trees are divided into two groups according to how they bear their seeds. Gymnosperms produce seeds that are often collected together into cones. Most conifers are gymnosperms. Angiosperms have flowers and produce their seeds inside a fruit. Deciduous trees are usually angiosperms.

In temperate regions, many deciduous trees drop their seeds in the autumn. Some seeds may be in nut form, like acorns and beechnuts, or they may have papery wings, like the seeds of sycamores. In tropical regions, seeds fall just before or during the dry periods.

ARE TREES IMMORTAL?

Trees have the ability to continue growing all their lives. This means that, in theory, they could live forever. When a tree dies, its death is usually caused by something in its environment, such as fire, wind, lightning, extremely dry periods, disease, insects, or being cut by humans. As a tree ages, it becomes more susceptible to disease and pests, and most trees die of several causes.

CLONING THE TREES

A clone is an exact genetic copy of its "parent," because it is grown from one of the parent's cells. Cloning may someday become an important part of the forest industry. Cells from a superior adult tree could be cloned in a laboratory and the resulting seedlings planted on tree farms. As the trees grew, they would resemble the parent tree and be more uniform than a natural forest. So far, cloned trees have been grown from cells of redwood, Douglas fir, apple, citrus, and poplar trees.

Common deciduous trees Typical deciduous trees in temperate regions include the ash, oak, maple, elm, poplar, birch, ginkgo, and magnolia. Tropical deciduous trees include the acacia, false beech, baobab, teak, casuarina, cannonball, ebony, peepul, sycamore, and rosewood.

OAK Oaks are perhaps the best-known deciduous trees in temperate regions. Oaks are very adaptable to different environmental conditions and are found in dry sandy plains to coastal swamps. Although there are many species, they all have at least one thing in common—the acorn, or nut, that bears a scaly cap. The leaves of some oaks, such as the red oak, turn a brilliant color in autumn. Others are less colorful. Oaks are the dominant trees of central and western Europe and are known for their excellent timber which is valued for furniture and hardwood floors.

ACACIA Acacia (uh-KAY-shuh) trees are called thorn trees in Africa, wattles in Australia, and mimosas in North America and Europe. They are found in dry tropical climates as they can tolerate long dry periods, but they do not grow very tall. Acacias usually have an umbrella shape, yellow flowers, and sharp spines. Their seeds are used by both humans and animals for food.

BAOBAB Found in dry tropical regions, the baobab (BA-oh-bab) has a soft, spongy trunk and lower branches for storing large quantities of water. During the dry seasons the trees lose their leaves, which reduces water loss.

Baobabs do not grow very tall—usually about 60 feet (18 meters) in height—but the trunk may reach 30 feet (9 meters) in diameter. Giant baobabs are usually several thousand years old. Many animals use the baobab for food and shelter, and pollination of its flowers is done by bats.

TEAK The teak is native to India, Burma, and Thailand, where it prefers well-drained soils. It is a large tree with a spreading crown and small white flowers, and mature trees may reach 150 feet (46 meters) in height.

THE WORLD'S OLDEST TREE SPECIES

The ginkgo (Ginkgo biloba), or maidenhair tree, is the only remaining survivor of a group of plants that thrived during the Permian period (260,000,000 years ago). Essentially unchanged since prehistoric times, the ginkgo is a smooth-barked tree with fan-shaped leaves and few branches. Although it forms seed cones, the ginkgo is deciduous, its leaves turning golden in fall and dropping.

In China, the ginkgo is sacred and is planted near temples. It is also popular in temperate climates of Europe and North America where it is grown as an ornamental. The seeds of the ginkgo are valued for their medicinal properties.

THE MIGHTY ACORN

It is an old saying that "Mighty oaks from little acorns grow." But acorns are appreciated for more than just turning themselves into oak trees. Many animals use them for food, and, at various times in history, humans have pounded them into a nutritious flour. Now a new use has been found: a cure for polluted rivers.

Acorns contain acornic acid, which binds with poisons from heavy metals, such as uranium, allowing these pollutants to be removed from rivers and streams. An estimated 2.2 pounds (4.8 kilograms) of acorns can clean up 3.5 tons (3.9 metric tons) of water.

Teak timber is known for its indestructibility. In places where they are sheltered from the weather, teak beams in Indian temples have been found to be in good condition after 1,000 years. Smaller pieces have survived at least 2,000 years in good condition.

ENDANGERED SPECIES

Trees can be threatened by natural dangers, such as forest fires, animals, and diseases, as well as by humans. Fires are more of a threat in dry climates, while animals and diseases seem prevalent in all climates. For example, when deer populations get too large, they can destroy forests by eating wildflowers and tree seedlings. The caterpillar of the gypsy moth eats the leaves from a variety of tree species. If enough of these insects attack a stand of trees, all the leaves are eaten and the trees die. Dutch elm disease, a fungus transmitted by the elm bark beetle that destroyed millions of elm trees in the United States, was reintroduced to England in logs exported from the United States. As a result, it is killing many elms there as well.

Pollution is also a threat to birch, beech, ash, and sugar maple trees in the northeastern United States. Pollution appears to weaken the trees so that pests and diseases overtake them more easily. In western Europe, the beech is in decline.

> ## THE FUNGUS THAT FELLED A GIANT
>
> Until 1940, the United States was home to the American chestnut tree, a graceful species with creamy white blossoms that often grew more than 100 feet (30 meters) tall. However, in 1904, foreign chestnut trees were imported into New York. These trees carried a fungus, called chestnut blight, that soon spread to American chestnuts. Over the next 40 years, the fungus destroyed almost all the American chestnuts in the eastern half of the country.
>
> Fortunately, the blight did not travel to the far west, and the American chestnut can still be found there. Also, the roots of the stricken chestnuts survived, because the blight could not reach them underground, and they continue to send up shoots. Unfortunately, the shoots soon die because the fungus is still in the soil. However, researchers are attempting to cross the American chestnut with other species that are resistant to the fungus in hopes of reestablishing the trees.

ANIMAL LIFE

From their roots to their tips, deciduous trees support a wide range of plant-eating animals and wildlife, while many other types of animals live among or beneath the trees. The animals can be classified as microorganisms, invertebrates, amphibians, reptiles, birds, and mammals.

MICROORGANISMS

A microorganism is an animal, such as a protozoa, that cannot be seen without the aid of a microscope. Every forest is host to millions of these tiny creatures. Microscopic roundworms, or nematodes, for example, live by the thousands in small areas of soil in deciduous forests and aid the process of decomposition.

BACTERIA

Bacteria are always present in woodland soil where they help decompose dead plant and animal matter. In temperate climates, bacteria help create nutrient-rich humus. Fewer bacteria are at work in dry climates or in moist climates with long dry seasons.

INVERTEBRATES

Animals without backbones are called invertebrates. They include simple animals, such as worms, and more complex animals, such as the wasp and the snail. Certain groups of invertebrates must spend part of their lives in water. Generally speaking, these types are not found in the trees, but in ponds, lakes, and streams, or in pools of rainwater.

Some invertebrates, such as beetles, are well adapted to life in dry tropical forests. They have an external skeleton, a hard shell made from a substance called chitin (KY-tin). Chitin is like armor and is usually waterproof,

A teak forest in Thailand. Overharvesting of teak trees has lead to their endangerment. (Reproduced by permission of Corbis. Photograph by Kevin R. Morris.)

protecting against the heat of the Sun and preventing its owner from drying out. These same invertebrates do not survive as well in temperate forests, because many adults die in the cold of winter. Their well-protected eggs or larvae may survive until the spring, however.

Food Many invertebrates eat plants or decaying animal matter. The larvae of insects, such as caterpillars, are the primary leaf eaters in the deciduous forest. The tent caterpillar, the larvae form of a moth, spins a large filmy tentlike cocoon around a tree branch, which it soon picks clean. Weevils drill holes in acorns which their larvae use for food, thus destroying the seed of the oak. Bees gather pollen and nectar (sweet liquid) from flowers, as do butterflies and moths, helping produce new plants through pollination. The arachnids (spiders), which are carnivores (meat eaters), prey on insects. Larger ones may even eat small lizards, mice, and birds.

Reproduction Most invertebrates have a four-part life cycle. The first stage of this cycle is spent as an egg. The egg's shell is usually tough and resistant to long dry spells in tropical climates. After a rain and during a period of plant growth, the egg hatches. The second stage is the larva (such as a caterpillar), which may actually be divided into several stages between which there is a shedding of the animal's outer skin. Larvae often spend their stage below the ground, where it is cooler and more moist than on the surface. The pupal, or third stage, is spent hibernating within a casing (such as a cocoon). When the animal emerges from this casing, it is an adult.

Common deciduous forest invertebrates
Invertebrates found in temperate forests include the tent caterpillar, the wooly bear caterpillar, the luna moth, the stag beetle, and the wolf spider. Other common invertebrates include the earthworm, the slug, the forest snail, and the acorn snail.

In tropical climates, many species of invertebrates, including the silkworm, inhabit moist forests. Dry climates have fewer insects, although termites and grasshoppers are common, as are many species of caterpillars.

AMPHIBIANS

Amphibians are vertebrates (animals with backbones) that usually spend part, if not most, of their lives in water. Frogs, toads, and salamanders

---

THE DESTRUCTIVE HITCHHIKER

In 1996, 2,400 trees in Brooklyn and Amityville, New York, were destroyed when the Asian long-horned beetle turned up in the United States. The 2-inch- (5-centimeter-) long insect lays its eggs just under the bark of maple, ash, elm, and horse chestnut trees, and when the larvae hatch and seek food, they cut off vital nutrients to the trees, starving them to death. It is believed that the beetle first made its way to this country by hitchhiking in the wood packing materials of goods shipped from China.

In 1998, this black and white beetle was discovered in Chicago, where 470 trees had to be cut down and burned in an effort to stop the insect's spread. Burning is done during the winter months while the beetle is inactive and cannot escape. Although it is hoped that all the infested trees have been identified and destroyed, no one can be sure that the beetles will not turn up somewhere else in the future.

live in significant numbers in temperate and moist tropical deciduous forests. Frogs and toads can also be found in dry tropical forests as long as there is a dependable source of water.

Because amphibians breathe through their skin, and only moist skin can absorb oxygen, they must usually remain close to a water source. Mating, egg-laying, and young-adulthood all take place in ponds, lakes, or pools of rainwater. Those that survive and reach maturity leave the pools for dry land where they feed on both plants and insects. In dry tropical forests, amphibians must find shade during the day or risk dying in the heat of the Sun.

Amphibians are cold-blooded animals, which means their bodies are about the same temperature as their environment. They, therefore, need a warm environment in order to be active. As temperatures get cooler, they slow down and seek shelter. In temperate climates, amphibians hibernate (remain inactive) during winter months. In hot, dry climates, amphibians go through estivation, a similar inactive period. While the soil is still moist from the rain, they dig themselves a foot or more into the ground, where they remain until the rains return. Only their nostrils remain open to the surface.

> **NOW YOU SEE IT, NOW YOU DON'T!**
> Many animals use camouflage (KAH-mah-flahj; protective coloration) to disguise themselves from predators. However, some predators use camouflage themselves to lurk among the leaves, waiting for lunch to pass by. The yellow color of the goldenrod spider, for example, enables it to hide in goldenrod flowers where it can sneak up on an insect sipping the flower's nectar.

Food Adult amphibians are usually carnivorous, feeding on insects, slugs, and worms. Salamanders that live in the water suck their prey into their mouths. Those that live on land have long, sticky tongues that capture food. Amphibian larvae are mostly herbivorous (plant eaters), feeding on vegetation. Frogs and toads are omnivores (both plant and meat eating) feeding on algae, plants, and insects such as mosquitoes.

Reproduction Mating and egg-laying for amphibians must take place in water. Male sperm are deposited in the water and must swim to the female's eggs in order to penetrate them. As the young develop into larvae and young adults, they often have gills for breathing.

Common deciduous forest amphibians Amphibians common in temperate forests include spring peepers, which are tiny frogs that climb trees; spotted newts; Fowler's toads; and marble salamanders. The African bullfrog is common to dry forests in Africa and the Pipid toad in South America. The flying frog is found in moist forests in India. It has a web of skin on both sides that begins at the wrist and attaches to the ankle. This skin acts as a parachute and enables the frogs to glide from one branch to another.

REPTILES

Reptiles that live in deciduous forests include many species of snakes and some lizards and turtles. The body temperature of reptiles changes with the temperature of the surrounding environment. Early in the day, they expose as much of their bodies as possible to the Sun for warmth. As temperatures climb, they begin to seek shade. During hot, dry periods, they must find shade or a hole in which to wait for cooler weather. During chilly nights, they become sluggish. In temperate climates, snakes may hibernate in burrows during the long winter.

Food The diet of lizards varies, depending upon the species. Some have long tongues with sticky tips and specialize in insects. Many are carnivores that eat small mammals and birds. The water they need is usually obtained from the food they eat.

All snakes are carnivores, and one good meal (such as a rabbit, rat, or bird) will last them for days or weeks. While constrictors squeeze their prey to death, other snakes kill their prey with venom (poison) injected through the snake's fangs.

Reproduction Reptile eggs are leathery and tough, and offspring are seldom cared for by the parents. Some females remain with the eggs, but most bury them in a hole then leave. The young are, therefore, left to hatch by themselves. Once free of the eggs, the babies dig themselves out of the hole and begin life on their own.

A calling spring peeper tree frog perched on a tree branch with an inflated vocal pouch. (Reproduced by permission of Corbis. Photograph by Joe McDonald.)

Common deciduous forest reptiles Temperate forest reptiles include the five-lined skink, eastern box turtle, and garter snake. Reptiles found in dry tropical forests include the night adder, the puff adder, the Gabonan adder, and the agama lizard. Moist tropical forests are home to the python and the calot lizard.

BIRDS

All forests have bird populations. Some species, such as the grosbeak, are migratory, which means they travel from one seasonal breeding place to another. During excessively cold or dry periods, most birds can simply fly to more comfortable regions. Others, such as the blue jay, prefer to stay in the same area year-round.

Feathers protect birds not only from cold winters but from tropical heat. Air trapped between layers of feathers acts as insulation against both climates.

Food Birds are found in greater variety and numbers where there is an ample supply of seeds, berries, and insects. Different birds seek food in different layers of the forest. Orioles and tanagers, for example, hunt for food high in the canopy. The white-breasted Nuthatche hunts for insects along the trunks of the trees, and towhees dig around on the ground. Some birds that live year-round in temperate climates such as chickadees and blue jays, may hide seeds in holes in trees where they can find them during the cold months. This hiding of seeds is called caching.

Reproduction In dry tropical climates, such as that in most of Australia, birds adapt their breeding habits to periods of rainfall, and breeding cycles may be far apart.

Although birds are free to fly away from uncomfortable temperatures during the rest of the year, the breeding cycle dictates when they can travel. They must remain in the same spot from the time nest building begins until the fledgling birds can fly, usually a period of many weeks.

Normally the parents sit on the nest to protect the eggs from heat or cold. During very hot weather, the parents may stand over the nest to give the eggs or the nestlings shade.

Common deciduous forest birds In temperate forests, common birds include screech owls, great horned owls, hummingbirds, woodpeckers, nuthatches, woodthrushes, American redstarts, hawks, blue jays, cardinals, scarlet tanagers, chickadees, and turkey vultures.

Wrens, falcons, weaverbirds, and chats are found in dry tropical forests. In the moist deciduous forests of Australia, currawongs build nests of sticks in the eucalyptus trees, and the thrush is common in India.

JAY More species of jays are found in the Northern Hemisphere; however, jays are also common in South America, Eurasia, and Africa. Their feathers may be drab or brightly colored, such as those of the North American blue jay. Many have crests and long tail feathers. They eat both plant matter and insects, and some species eat the eggs of other birds.

Jays are known for their bold, aggressive behavior and loud, harsh voices. Gangs of jays often harass other birds and even humans. The bluejay was immortalized in Mark Twain's famous story, "What Stumped the Bluejays."

FALCON Found all over the world, falcons are birds of prey. They are characterized by long, pointed wings and the ability to fly very fast. Some species fly to a high altitude and then "dive" on their prey, killing it with the powerful blow of their clenched talons. A falcon's power dive may reach 180 miles (288 kilometers) per hour.

Since about 722 B.C., falcons have been trained to hunt specific animals, such as rabbits and other birds. The sport of falconry is still practiced and is popular in Saudi Arabia, India, and Pakistan.

A blue jay perched on a tree branch. Jays are one of the more common birds found in deciduous forests. (Reproduced by permission of Corbis. Photograph by Richard Hamilton Smith.)

MAMMALS

Mammals are vertebrates that are covered with at least some hair and bear live young. Although only a few large mammals, such as bear and deer, live in temperate deciduous forests, many small ones, including mice, squirrels, woodchucks, and foxes, make their homes there. During the cold winters, many mammals burrow underground or find some other kind of shelter. Squirrels, for example, build nests high in the forest canopy. Some mammals, like bears, hibernate. In dry tropical climates, small mammals, such as rodents, may estivate during the dry season.

Food Some mammals, such as mice, eat plants and insects. Others, like squirrels and hedgehogs, will eat bird and reptile eggs and young. Many smaller mammals do not need to drink water as often because they obtain moisture from the food they eat.

Reproduction The young of mammals develop inside the mother's body. In this way they are protected from heat, cold, and predators. Mammals also produce milk to feed their young, and those that live in dens must remain nearby until the young can survive on their own. In temperate forests, the young are usually born in the spring so they have many warm months ahead in which to grow strong.

Common deciduous forest mammals Temperate forests are home to shrews, woodmice, grey foxes, chipmunks, ground squirrels, badgers, black bears, silver-haired bats, raccoons, opossums, weasels, cottontail rabbits, grey squirrels, skunks, flying squirrels, and bobcats.

Dry deciduous forests support grazing animals, such as zebras and gazelles, which feed on the grasses that grow beneath the widely spaced trees. In the moist deciduous forests of Australia, red and gray kangaroos browse on leafy foliage. Wombats, opossums, and koalas live among the trees. Tigers are common in the moist forests of Asia, as are elephants and buffalo.

RACCOON Raccoons are medium-sized furry mammals with stout bodies, short legs, and long, bushy tails. They are perhaps best known for the black "mask" around their eyes and their proficiency as burglars. They prefer living in the woods where they hunt rodents, birds and bird eggs, berries, fruit, and other plant matter. However, they also love corn and melons and other niceties of civilized life, which makes them expert raiders of suburban gardens and garbage cans.

Raccoons live in dens made in hollow trees. This is where they spend the daylight hours, and they go out to hunt at night. Although they sleep more during cold seasons, they do not hibernate. Intelligent and curious animals, raccoons are hunted for their fur and meat.

TIGER The tiger is one of the largest and strongest carnivores. Indian tigers may grow to 10 feet (3 meters) in length and weigh as much as 600 pounds (275 kilograms). They are found only in Asia where they prefer humid forests, the dense undergrowth of which allows them to sneak up on their prey.

Around 1800, there were as many as 50,000 tigers in India alone. However, during the nineteenth century, tiger hunting became very popular. Since then, the tiger has completely disappeared from Turkey and parts of China and Siberia. India now has only about 3,000.

Although accounts of tigers killing humans are rare, the animals are extremely dangerous. Between 1973 and 1982 in a single Indian nature reserve, tigers killed more than 100 people, including many children.

ENDANGERED SPECIES

In the United States, acid rain (a mixture of water vapor and polluting compounds in the atmosphere that falls to the Earth as rain or snow) has

---

THE WEB OF LIFE: NATURE'S JOHNNY APPLESEEDS

Johnny Appleseed was the nickname of John Chapman (1774–1845) a man who traveled around North America planting apple trees. Some animals fulfill similar roles, although perhaps not on purpose. Squirrels and chipmunks, for example, in preparation for the long winter, bury acorns and other nuts in the ground, where they can dig them up later and eat them. However, they often forget where they buried them, and, in the spring, the nuts sprout and send up shoots. Thanks to the poor memories of squirrels and chipmunks, forests spread and grow.

---

endangered the peregrine falcon. In Asia, South America, and Africa, all the big cats, including tigers, leopards, and cheetahs, are endangered as humans encroach on their habitats. In the United States and Canada, the timber wolf is threatened; in Australia, the koala and some species of kangaroos; and in Africa and Asia, the elephant. In some areas the animals are overhunted; in others their habitats are disappearing.

A raccoon opens its mouth in defense as it looks out from the hollowed knot of a snow-covered tree near Boulder, Colorado. (Reproduced by permission of Corbis. Photograph by W. Perry Conway.)

HUMAN LIFE

Without forests, there would probably be no human life on Earth. Many animals, including humans, are creatures of the forest. Until humans learned to hunt, they ate plant foods such as bark, nuts, and berries. The earliest records of humankind show that humans and the great forests evolved together. As humans learned to hunt animals for food and clothing, they sought out the forests, which was plentiful in animal life. Humans and forests will always be connected this way.

IMPACT OF THE DECIDUOUS FOREST ON HUMAN LIFE

Forests have an important impact on the environment as a whole. From the earliest times, forests have also offered food and shelter, a place to hide from predators, and many useful products.

Environmental cycles Trees, soil, animals, and other plants all interact to create a balance in the environment from which humans benefit. This balance is maintained in what can be described as cycles.

THE OXYGEN CYCLE Plants and animals take in oxygen from the air and use it for their life processes. When animals and humans breathe, the oxygen they inhale is converted to carbon dioxide, which they exhale. This oxygen must be replaced, or life could not continue. Trees help replace oxygen during photosynthesis, when they release oxygen into the atmosphere through their leaves. A global research project to measure the overall influence of forests on the Earth's atmospheric balance is underway.

THE CARBON CYCLE Carbon dioxide is also necessary to life, but too much is harmful. During photosynthesis, trees and other plants pull carbon dioxide from the air. By doing so they help to maintain the oxygen/carbon dioxide balance in the atmosphere.

A Siberian tiger in captivity. Tigers are no longer found in the wild in Siberia. (Reproduced by permission of Field Mark Publications. Photograph by Robert J. Huffman.)

When trees die, the carbon in their tissues is returned to the soil. If decaying trees become part of the Earth's crust, after millions of years, this carbon becomes oil and natural gas.

THE WATER CYCLE Forests shade the snow, allowing it to remain in deep drifts. Their root systems and fallen leaves help build an absorbent covering on the ground, allowing rain water and melting snow to soak into the soil and trickle down to feed underground streams and groundwater supplies.

Not only do forests help preserve water in this way, but they also protect the land. Trees act as barriers, or walls, that help to reduce the strong forces of winds and rain. When forests are cut down, this barrier is removed and the topsoil is either blown away, or during heavy rains, the soil washes away. As a result, flooding is more common because there is no absorbent layer to soak up the rain. For example, since 1997, parts of India and Bangladesh have suffered severe flooding caused in part by the cutting of the forests in the nearby Himalaya Mountains.

Trees take up water through their roots and use it for their own life processes. Extra moisture is then released through their leaves back into the atmosphere, where it forms clouds and once again falls as rain or snow, helping the water cycle to continue.

THE NUTRIENT CYCLE Trees get the mineral nutrients they need from the soil. Dissolved minerals are absorbed from the soil by the tree's roots and are sent upward throughout the tree. These mineral nutrients are then used by the tree much like humans take vitamins. When the tree dies, these nutrients, which are still contained within parts of the tree, decompose and are returned to the soil. As a result, these nutrients are available for other plants and animals to use.

FOOD

Since the earliest times, forests have been the home of game animals, such as deer, which have supplied meat for hunters and their families. Forests also supply fruits, nuts, seeds, and berries, as well as vegetation for livestock and the honey of bees. Cranberries, gooseberries, strawberries, raspberries, huckleberries, and currants, all referred to as brambles, grow in temperate woodlands.

SHELTER

During prehistoric times, humans lived in the forest because it offered protection from the weather and dangerous animals. Today, people of developed countries who choose to live in forested areas usually do so because they enjoy their beauty. However, some native tribes still live in forests, as their ancestors did many years ago.

ECONOMIC VALUES

Forests are also important to the world economy. Many products used commercially are obtained from forests, such as wood, medicine, tannis, and dyes.

Wood Trees produce one of two general types of wood, hardwood or softwood, based on the trees' cell wall structure. Hardwoods are usually produced by deciduous trees, such as oaks and elms. Most coniferous trees,

COMMON HARDWOODS

North America and Europe	Central and South America	Africa	Asia	Australia
Alder	Aromata	African oak	Aralia	Black bean
Ash	Balsa	African walnut	Bow wood	Cedar
Basswood	Bois gris	Camphorwood	Boxwood	Coconut palm
Beech	Bois lait	Canarium	Canarium	Gum
Birch	Brazilwood	Ebony	Cedar	Ironbark
Boxwood	Chilean laurel	Mahogany	Cinnamon	Peppermint
Cherry	Greenheart	Olive	Ebony	Silky oak
Chestnut	Lancewood	Teak	Elm	Tasmanian oak
Dogwood	Lignum vitae		Gum	Tea tree
Elm	Mahogany		Horse chestnut	Turpentine
Hickory	Pepper		Indian laurel	Walnut
Holly	Rosewood		Japanese alder	
Hornbeam	Satinwood		Japanese ash	
Lime	Snakewood		Japanese birch	
Magnolia	Teak		Japanese maple	
Maple	Tulipwood		Katsura	
Oak	Yokewood		Rosewood	
Olive			Sandalwood	
Pear			Teak	
Plane			Tree of Heaven	
Poplar			Walnut	
Sycamore			Willow	
Walnut				
Willow				

such as pines, produce softwoods. However, these names can be confusing, because some softwood trees, such as the yew, produce woods that are harder than many hardwoods. And some hardwoods, such as balsa, are softer than most softwoods.

Wood is used for fuel, building structures, and manufacturing other products, such as furniture and paper. Wood used for general construction is usually softwood. In order to conserve trees and reduce costs, some manufacturers have created engineered wood, which is composed of particles of several types of wood mixed with strong glues and preservatives. Engineered woods are very strong and can be used for many construction needs.

Hardwood from deciduous trees is more expensive because the trees grow more slowly and, as a result, it is used primarily for fine furniture and paneling. Most of this hardwood comes from forests in Europe. In the United States, the major commercial hardwoods are sugar maple, red oak, black cherry, yellow birch, white ash, black walnut, and beech.

Farmland In temperate regions, the soil beneath a deciduous forest is usually extremely fertile. For this reason, forest land has often been cleared for agricultural purposes. As the United States was settled, most of the original deciduous forest in the East was cut down to accommodate farms and large plantations. After the land was exhausted and farming became centralized in the plains of the Midwest, the forests gradually grew back. Oaks, maples, and hickories now flourish in the same regions that supported their ancestors 400 years ago.

Medicines Since the earliest times, plants have been used for their healing properties. The ancient Greeks, for example, used extracts from willow bark

WORLD TRADE IN TIMBER PRODUCTS (PARTIAL LIST)

Country	Exports: tons (metric tons)	Imports: tons (metric tons)
United States	12,948 (11,744)	16,778 (15,218)
Canada	17,588 (15,952)	2,938 (2,665)
Great Britain	—	5,484 (4,974)
Russia	16,050 (14,577)	—
China	554 (493)	4,348 (3,944)
Japan	—	31,679 (28,733)
Indonesia	13,961 (12,663)	—
Sweden	4,403 (3,994)	2,023 (1,835)

to relieve pain, as did certain tribes of Native Americans. During the nineteenth century, scientists finally discovered what the pain-killing ingredient was and gave it a name—salicylic acid. We are still familiar with it today by its commercial name, aspirin. Many drug companies maintain large tracts of forest as part of their research programs.

Tannins and dyes Tannins are chemical substances found in the bark, roots, seeds, and leaves of many plants. These tannins can be extracted by boiling or soaking the plant material. The extract is then used to cure leather, making it soft and supple. Trees used for tannins include the oak, the chestnut, and the quebracho.

Dyes used to color fabrics can be obtained from oak, alder, birch, walnut, brazilwood, and logwood. However, natural dyes are no longer used much commercially, since dyes can be produced more cheaply from chemicals.

Recreation More people live in cities today than ever before, and many people feel the need to escape to more natural surroundings occasionally. The beauty and quiet of deciduous forests draw many visitors for hiking, horseback riding, skiing, fishing, hunting, birdwatching, or just sitting and listening to nature.

HEALING SAPS AND GUMS

Gums and saps from certain trees are very useful commercially. For example, the sap of the spiny acacia, called gum arabic, was first used by the ancient Egyptians in making inks, has been used in the past in medicines and adhesives, and is still used in making watercolor paints. But these helpful liquids, which usually ooze from wounds in a tree, are produced only by trees that are unhealthy because of dry weather or poor soil and have been infected by microorganisms or fungi. The sap or gum may actually help the tree heal itself.

IMPACT OF HUMAN LIFE ON THE DECIDUOUS FOREST

Just as the forest has had an effect on human life, human life has had an effect on the forest. About 10,000 years ago, forests covered about half of the Earth's land surface. Today, they cover less than one-third. Nearly 2,000,000,000 tons (1,814,000,000 metric tons) of timber are cut from the world's forests each year. Most of the losses in forest cover have occurred in developing nations where wood is used for fuel and where trees are cleared away for farming. The rest of the crop is used commercially.

Use of plants and animals During the Middle Ages (500–1450) in Europe and the eighteenth and nineteenth centuries in America, it was thought that the forests were indestructible. They seemed to go on forever. Eventually, however, the cutting of trees brought an end to many ancient forests. Only when they were left alone to regrow or were deliberately replanted, did the forests begin to recover.

In the United States, environmentalists are in disagreement with logging companies over the cutting of old-growth forests. Although trees are a renewable resource, old-growth trees may take more than 100 years to be replaced. Oftentimes, the logging companies replace the old trees with seedlings from other species that grow faster, and the original species are never replanted. Or the companies may replace only certain species, so that the forest's diversity is destroyed, along with plants and animals that depend on it.

Other threats to North American forests include mining operations that want to locate on public forest lands, and the cutting down of trees in order to develop the land for homes and businesses.

Conservation laws in many states protect trees on public lands, while other laws protect forest creatures from being overhunted.

In tropical regions, much forest land is being lost as populations grow and want the land for farms. "Slash and burn" agriculture is practiced in which the trees are cut and burned. The land is then used between one to five years for farming. When the land, which is usually poor, will no longer support crops, it is abandoned, and another forest plot is cut down somewhere else. As a result, many animals and plants are losing their habitats.

Quality of the environment Destruction of the forests does not mean just loss of their beauty and the products they provide. Soil quality also declines. Some soil is washed away by rain, and the loss of plant life means the soil will not be built back up again.

Water quality and supply also suffers. Because the trees are gone, rain does not seep into the soil, and underground water reserves are not replaced. Soil that is washed away often ends up in streams and rivers, and if the quantity of soil is large enough, fish may die.

Air quality is also reduced when forests are destroyed. Not only do trees put oxygen back into the air, but soot and dust collect on their leaves. When it rains, this now trapped soot and dust are washed to the ground where they enter the soil. In this way, trees help keep the air clean. However, when the trees are cut down, the dust and soot remain in the air as air pollution continually floating around.

With the popularity of the automobile, carbon dioxide and other undesirable gases have built up in the atmosphere. Some scientists believe that these gases are helping to raise the temperature of the Earth's climate by forming an invisible layer in the atmosphere which keeps the heat in instead of letting it escape into the upper atmosphere. This is called the "greenhouse effect." Because forests help remove carbon dioxide from the air, cutting them down may be helping cause this global warming. If the Earth continues to grow warmer, many species of plants and animals could become extinct.

Air pollution is also helping to destroy forests, some of which take on an off-color, or sickly, appearance in polluted environments. Air pollution also causes acid rain, which damages some types of trees. Acid rain occurs when certain compounds in polluted air mix with water vapor and fall to the Earth when it rains. When acid rain is absorbed into the soil, it can destroy nutrients and make the soil too acidic to support some species of trees. Forests in northern Europe, southern Canada, and the eastern United States have been damaged by acid rain.

Off-road vehicles, such as motorbikes and four-wheelers, tear up forest undergrowth, kill flowers, scare away animals, and destroy the peace of the forest.

Forest management The National Forest Service was established in the United States to protect forest resources. More than 190,000,000 acres (76,760,000 hectares) of land are now publicly owned. Most of this acreage is west of the Mississippi. Forests in the eastern half of the United States are managed by state programs. Most of these forests consist of land that can be used for logging and other commercial purposes; however, certain portions are kept for recreation and conservation.

Many other nations, including Great Britain, Japan, China, and India, have also established programs to conserve and replant forests.

NATIVE PEOPLES

By 1950, native peoples in industrialized parts of the world had abandoned most of their tribal lands and customs. Many went to live in cities. However, some still live a traditional lifestyle today.

The Birhor The Birhor of central India are one example. They live in the forests as nomadic hunter-gatherers and supplement their diet with rice for which they trade forest products such as bark fiber rope.

Algonquin and Iroquois Tribes Until the European settlers forced them off their ancestral lands, many Native American tribes lived in the deciduous woodlands of the northeastern United States. They included tribes who spoke the Iroquois language—the Cayuga, Erie, Huron, Mohawk, Oneida, Onondaga, Seneca, Tuscarora, and Neutral; and those who spoke Algonquin—the Delaware, Fox, Illinois, Kickapoo, Mohican, Massachuset, Menominee, Miami, Mohegan, Ottawa, Pequot, Sauk, Shawnee, Shinnecock, and Wampanoag.

Long winters in those regions limited farming, so that most foods were collected from the wild. These included fish, game, maple syrup, and wild rice.

This dead spruce and dead fir trees in the forests of the Great Smoky Mountains are the result of acid rain. (Reproduced by permission of JLM Visuals.)

Nearly all the native groups who survived contact with European settlers were forced to move to reservations in Oklahoma during the early nineteenth century.

Ainu The Ainu are native peoples who live on Hokkaido, the northernmost island of Japan, and on the Russian island of Sakhalin in the North Pacific. Originally hunters and gatherers, the men used the bow and arrow to hunt bear, deer, fox, otter, and other animals during the winter; in summer they fished. The women gathered roots, berries, mushrooms, and nuts and also did some farming. The men were skilled woodcarvers, and the women wove fabrics and did embroidery. Their chief musical instruments were the drum and flute.

At one time, the Ainu lived throughout the Japanese islands, but gradually, as the Japanese population moved in and expanded, the Ainu have been pushed to their present territory. They have also intermarried with the Japanese, and only about 25,000 Ainu of unmixed descent remain.

THE FOOD WEB

The transfer of energy from organism to organism forms a series called a food chain. All the possible feeding relationships that exist in a biome make up its food web. In the deciduous forest, as elsewhere, the food web consists of producers, consumers, and decomposers. An analysis of the food web shows how energy is transferred within the biome.

Green plants are the primary producers in the forest. They produce organic materials from inorganic chemicals and outside sources of energy, primarily the Sun. Trees and other plants turn energy into plant food.

Animals are consumers. Plant-eating animals, such as grasshoppers and mice, are the primary consumers in the forest food web. Secondary consumers eat the plant-eaters. Tertiary consumers are the predators, like owls, foxes, and tigers. They are carnivores. Humans fall into this category. Humans are also omnivores, which means they eat both plants and animals.

Decomposers feed on dead organic matter, and include plants like fungi and animals like the turkey vulture. In moist environments, bacteria also help in decomposition. When fallen leaves from deciduous trees carpet the ground forming a thick layer, bacteria feed on the leaves and help decompose them into humus.

SPOTLIGHT ON DECIDUOUS FORESTS

FORESTS OF MEDIEVAL EUROPE

About 5,000 years ago, birch trees dominated the forests of the northern hemisphere. As the climate gradually became warmer, however, other trees began to take over. By Medieval times (500–1450), many hardwood trees, such as beeches and long-lived oaks, ranged over much of central and western Europe north of the Alps and the Pyrenees Mountains, and eastward across Russia to the Ural Mountains. Oaks and beeches usually formed the canopy, with maples and birches in the secondary layer, and dogwoods, hawthorns, and hollies closer to the ground.

The forests were home to falcons, hawks, herons, owls, deer, wolves, boars, otters, squirrels, foxes, badgers, and other wild animals. Many of these were hunted, some for food and others to protect crops and domesticated animals.

For people living there, the forests were a source of timber for building houses, vines and leaves for feeding livestock, and game animals for hunting. Bees, prized for their honey, were often kept within the forest, and the fruits and nuts of many trees, such as the hazel, were used as food. Pigs, for example, were kept in the forest, where they could feed on acorns. Valued for these nuts, oaks were allowed to grow old. Wood was prized as fuel, not only as logs but as charcoal, which is created when wood is partially burned. Charcoal produces a very hot fire and was used in beer brewing and forging iron. A by-product of burning wood, wood ash, was needed for making glass and soap. Wood was also used to build boats, carts, furniture, and even shoes.

> **FORESTS OF MEDIEVAL EUROPE**
> **Location:** Northern and western Europe
> **Classification:** Temperate

Open land was used for farming, but as populations grew, more and more forests were cut down and used for planting crops. By the eleventh century, the heavy plow came into use. It worked the soil more efficiently than previous methods and increased the amount of land that could be used for agriculture.

By 1300, great expanses of forests in Europe had been destroyed. In France, only 37,000,000 acres (14,984,000 hectares) of forest remained, and wood was so scarce in northern France that peasants could not afford wooden coffins. As a result, the coffins were rented for the ceremony, and afterward the undertaker would dump the corpse into the grave, keeping the coffin for another use. Eventually, in some regions such as Germany, restrictions were imposed that limited the number of trees that could be cut and made the use of live trees for fuel illegal. The practice of hunting also helped protect the forests because rich landowners wanted the game animals that lived in them to survive.

DECIDUOUS FORESTS OF JAPAN

Few countries have as large a percentage of forested land as Japan. Japan is a series of islands, and deciduous forests cover at least half of the largest island, Honshu, and the lower-lying parts of southern Hokkaido.

Beeches are the dominant trees, but many other species are mixed in, including oaks, chestnuts, maples, and limes. Oaks and chestnuts are also cultivated on privately owned tree plantations. Because Japan is such a small country in terms of available land, much of that land is intensely cultivated. This has led to the growing of miniature trees that can be planted in small gardens.

> **DECIDUOUS FORESTS OF JAPAN**
>
> **Location:** Northern Japan
>
> **Classification:** Temperate

The climate of Japan's deciduous forests is temperate but cool. Rain is plentiful, as is snow in winter. The stately beeches, which have bright green leaves in summer, turn a rich gold-brown in autumn.

Few trees are planted in Japan for timber. Most commercially grown trees, such as the beautiful Japanese maple and the flowering cherry, are instead sold for ornamental purposes.

DECIDUOUS FORESTS OF CHINA

China is the third largest country in the world, and its forested land stretches over many hundreds of thousands of square miles (square kilometers). In the north, the climate is cool temperate. In the northeast, oak, ash, birch, and poplar forests predominate. In the south, the climate is warm temperate. No single species predominates in the southeast, and even the canopy shows a mixture of trees, including oak, maple, poplar, boxwood, and sweetgum.

> **DECIDUOUS FORESTS OF CHINA**
>
> **Location:** Eastern China
>
> **Classification:** Temperate

Over the centuries, most of China's forests were destroyed by war, for timber and fuel, and to make way for agriculture. By 1949, only 8 percent of the land remained forested. However, after considerable reforestation efforts, by 1980 another 4 percent was added—an area larger than all of Great Britain. Residents are encouraged to plant trees around their homes, and in northern China, an 1,800-mile- (2,880-kilometer-) long belt of trees has been planted as a windbreak against the icy winds blowing in from northern deserts.

DECIDUOUS FORESTS OF NORTH AMERICA

Deciduous forests are found in North America in portions of southern Canada, New England, the upper Midwest, the Appalachian region, along the Mississippi, and in the southeastern states.

When European settlers arrived in America in the 1600s, these forests were thick with oak, beech, and chestnut trees. Many trees were at least 100 feet (31 meters) tall and shaded a wide variety of other plants. Wild turkeys, passenger pigeons, moose, cougars, bison, beavers, otters, bears, and wolves were plentiful. By 1970, however, most of the deciduous forests had been cut down, sacrificed to the building of houses and the need for farmland. Only in remote areas where the land was too steep to allow easy access did the original forest survive.

In some areas, deciduous forests are recovering. In New England, for example, there is more forest now than there was 100 years ago. There, deciduous trees, such as sugar maple, birch, beech, and hemlock, mingle with coniferous forests. From Minnesota and Michigan eastward toward New England the land is now covered in pine forests that took over during a severe dry period during the 1500s. Since then, the deciduous trees have recovered, and the area is being reclaimed by oak, beech, maple, aspen, and other species. In the southern and southeastern United States the land was once covered primarily by deciduous trees, such as oaks, which were cut for timber and to clear the land for farms. Now, coniferous trees predominate. If the land remains undisturbed, however, deciduous trees will once again take over.

> **DECIDUOUS FORESTS OF NORTH AMERICA**
>
> **Location:** Southern Canada and eastern United States
>
> **Classification:** Temperate

DECIDUOUS FORESTS OF AUSTRALIA

Only about 15 percent of the total area of Australia has enough rainfall to support forests. Its central region is primarily desert mixed with grassland, and its forests grow in bands on the outer fringes of the continent. Although most of these trees are evergreen, moist deciduous forests can be found along the north and northeast coasts. Along the southern edge of the continent are open woodlands and dry tropical forest mixed with grasslands.

In the more humid upland and coastal areas of southwestern and eastern Australia, eucalyptus, angophora, and sheoaks predominate, mixed with beeches in some areas and shrubs having leathery leaves. In the south, acacia, or wattle, trees are common.

> **DECIDUOUS FORESTS OF AUSTRALIA**
>
> **Location:** Along the coastlines of the continent
>
> **Classification:** Moist tropical and dry tropical

Woodland animals are varied and include many poisonous snakes. One interesting species, the carpet snake, is not poisonous but suffocates its victims by, wrapping itself around them. A climber, the carpet snake is often turned loose in barns by farmers who want it to catch rats and mice.

Forest birds include budgerigars, currawongs, honey-eaters, and laughing kookaburras, which are immortalized in the song "Laugh, Kookaburra, Laugh."

Mammals are predominantly marsupials, such as wombats and kangaroos, that carry their young in a pouch. Kangaroos range through the open woodlands of eastern Australia.

The native peoples of Australia are the Aborigines, who live in dry regions. A more complete discussion of the Aborigines can be found in the chapter titled "Desert."

DECIDUOUS FORESTS OF CENTRAL INDIA

Central India, in the general area between the cities of Delhi and Nagpur, is a dry tropical region with dry and mixed deciduous forest. Trees found here include teak, mango, mohwa, jamun, gardenia, sal, and palm. Much of the area is grassland, and bamboo is common.

> **DECIDUOUS FORESTS OF CENTRAL INDIA**
>
> **Location:** Central India between Delhi and Nagpur
>
> **Classification:** Dry tropical

The region supports many parks and nature preserves, including Kanha Tiger Reserve and Taroba and Shivpuri National Parks.

Animals in this region include jays, peacocks, demoiselle cranes, wild boar, sambal and chital deer, gazelles, jackals, leopards, and tigers.

Forestry accounts for only about 1.7 percent of India's income, and forests are not accessible for commercial development. The government is attempting to increase the forested area to 33.3 percent and improve lumber output.

DECIDUOUS FORESTS OF SOUTHERN INDIA

Several regions at the southern tip of India support moist tropical deciduous forest and mixed forests with many evergreens. Trees include teak, ebony, and bejal. Except for areas devoted to the cultivation of teak, undergrowth tends to be thick. Open areas are covered by grasses and bamboo.

> **DECIDUOUS FORESTS OF SOUTHERN INDIA**
>
> **Location:** The tip of India, south of Bangalore
>
> **Classification:** Moist tropical

Several reserves and national parks are located here, including the Mudumalai Sanctuary, the Nagarhole Sanctuary, and the Bandipur Tiger Reserve.

The animal population is rich and varied. Termite nests can be seen along the roadways. Pythons, cobras, and monitor lizards are common. Birds include peacocks, parrots, pigeons, hornbills, and drongos of Paradise. Giant squirrels, elephants, gaurs, lipped bears, leopards, and tigers also make their homes in the forest.

DECIDUOUS FORESTS OF TANZANIA AND KENYA

In Tanzania, the forest is called *miombo*. Trees are short—rarely over 50 feet (15 meters) tall with flat tops. During the dry season, which may last as long as seven months, the trees shed their leaves. Grasses and other plants die off, and the forest looks brown and scorched. When new leaves begin to appear, they signal the start of the rains, which give new life to the region.

Tree seedlings often develop large tap roots used to store water. During this time there is little development of the tree above the ground. Trees may typically take as long as seven years to grow more than 1 foot (30 centimeters) in height.

> ## DECIDUOUS FORESTS OF TANZANIA AND KENYA
> **Location:** Southeastern Africa
> **Classification:** Dry tropical

In Kenya, acacia woodlands are common, although the baobab is predominant. Seasonal droughts cause vegetation to die. Many trees store water in tap roots or, like the baobab, in their trunks. A single baobab trunk can contain up to 25,000 gallons (94,625 liters) of water.

Common animals include the zebra and wildebeest. Lions are the largest predator.

FOR MORE INFORMATION

BOOKS

Cricher, John C. *Field Guide to Eastern Forests*. Peterson Field Guides. Boston: Houghton Mifflin, 1988.

Nardi, James B. *Once Upon a Tree: Life from Treetop to Root Tips*. Iowa City, IA: Iowa State University Press, 1993.

Schoonmaker, Peter S. *The Living Forest*. New York: Enslow, 1990.

ORGANIZATIONS

Center for Environmental Education
  1725 De Sales Street NW, Suite 500
  Washington, DC 20036

Environmental Defense Fund
  257 Park Ave. South
  New York, NY 10010
  Phone: 800-684-3322; Fax: 212-505-2375
  Internet: http://www.edf.org

Environmental Network
  4618 Henry Street
  Pittsburgh, PA 15213
  Internet: http://www.envirolink.org

Environmental Protection Agency
  401 M Street, SW
  Washington, DC 20460
  Phone: 202-260-2090
  Internet: http://www.epa.gov

Forest Watch
  The Wilderness Society
  900 17th st. NW
  Washington, DC 20006
  Phone: 202-833-2300; Fax: 202-429-3958
  Internet: http://www.wilderness.org

Friends of the Earth
  1025 Vermont Ave. NW, Ste. 300
  Washington, DC 20003
  Phone: 202-783-7400; Fax: 202-783-0444

Global ReLeaf, American Forests
  PO Box 2000
  Washington, DC 20005
  Phone: 800-368-5748; Fax: 202-955-4588
  Internet: http://www.amfor.org

Greenpeace USA
  1436 U Street NW
  Washington, DC 20009
  Phone: 202-462-1177; Fax: 202-462-4507
  Internet: http://www.greenpeaceusa.org

Isaak Walton League of America
  SOS Program
  1401 Wilson Blvd., Level B
  Arlington, VA 22209

Sierra Club
  85 2nd Street, 2nd fl.
  San Francisco, CA 94105
  Phone: 415-977-5500; Fax: 415-977-5799
  Internet: http://www.sierraclub.org

World Wildlife Fund
  1250 24th Street NW
  Washington, DC 20037
  Phone: 202-293-4800; Fax: 202-293-9211
  Internet: http://www.wwf.org

WEBSITES

Note: Website addresses are frequently subject to change.

National Geographic Magazine: http://www.nationalgeographic.com

National Park Service: http://www.nps.gov

Nature Conservancy: http://www.tnc.org

Scientific American Magazine: http://www.scientificamerican.com

BIBLIOGRAPHY

Cole, Wendy. "Naked City: How an Alien Ate the Shade." *Time.* (February 15, 1999): 6.

Dixon, Dougal. *Forests.* New York: Franklin Watts, 1984.

"Forests and Forestry." *Grolier Multimedia Encyclopedia.* New York: Grolier, Inc., 1995.

Ganeri, Anita. *Forests.* Habitats. Austin, TX: Raintree Steck-Vaughn, 1997.

International Book of the Forest. New York: Simon and Schuster, 1981.

Kaplan, Elizabeth. *Temperate Forest.* Biomes of the World. Tarrytown, NY: Benchmark Books, 1996.

Masso, Renata. *India.* World Nature Encyclopedia. Milwaukee, WI: Raintree Steck-Vaughn, 1989.

Sayer, April Pulley. *Temperate Deciduous Forest.* New York: Twenty-First Century Books, 1994.

Silver, Donald M. *One Small Square: Woods.* New York: McGraw-Hill, 1995.

"Tree." *Encyclopaedia Britannica.* Chicago: Encyclopaedia Britannica, Inc., 1993.

DESERT

All deserts have two things in common: they are dry, and they support little plant and animal life. If a region receives an average of less than 10 inches (25 centimeters) of rain each year, scientists classify it as a desert. Contrary to what most people believe, all deserts are not hot. Some ice deserts near the North and South Poles are so cold that all moisture is frozen. This chapter will discuss deserts in tropical and temperate areas. (Tropical areas are those near the equator. Temperate areas are between the equator and the North and South Poles).

True deserts cover about 14 percent of the world's land area, or about 8,000,000 square miles (20,800,000 square kilometers). Another 15 percent of the Earth's land surface possesses some desertlike characteristics. Most deserts lie near the tropic of Cancer and the tropic of Capricorn, two lines of latitude lying about 25 degrees from the equator (see map). The area between these two lines is called the Torrid Zone. ("Torrid" means very hot.)

HOW DESERTS ARE FORMED

In general, deserts are caused by the presence of dry air. The average humidity (moisture in the air) is between 10 and 20 percent. In some cases, mountain ranges prevent moisture-laden clouds from reaching the area. Mountains can also cause heavy, moisture-filled clouds to rise into the colder atmosphere. There, the moisture condenses and falls in the form of rain, leaving the air devoid of moisture as it crosses the range. In other cases, certain wind patterns along the equator bring air in from dry regions. In another scenario, cold-water ocean currents can cause moist air to drop its moisture over the ocean. The resulting dry air quickly evaporates (dries up) ground moisture along the coastal regions as it moves inland.

Deserts have always existed, even when glaciers covered large portions of the Earth during the great Ice Ages. Although geological evidence is scarce,

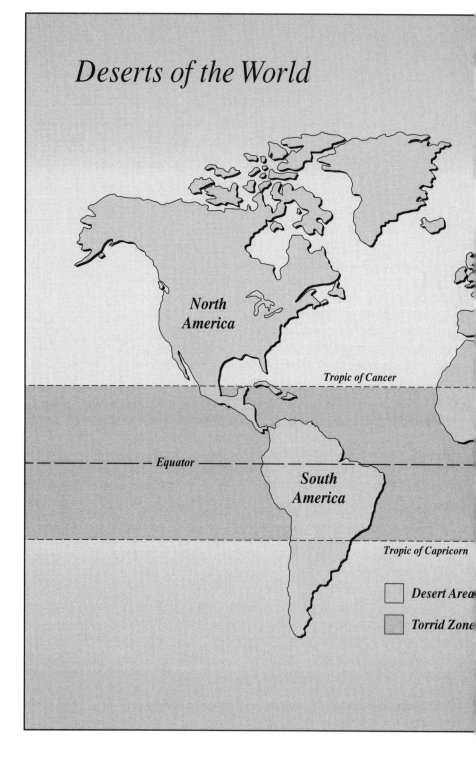

Deserts of the World

North America

Tropic of Cancer

Equator

South America

Tropic of Capricorn

Desert Area

Torrid Zone

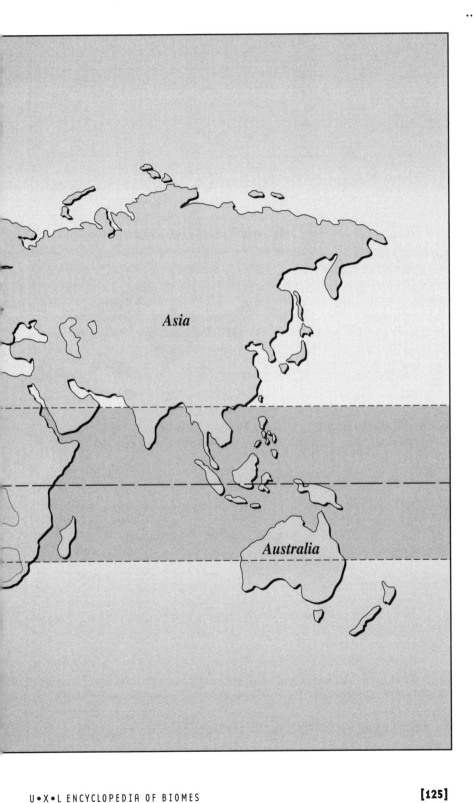

scientists tend to agree that some desert areas have always been present, although they were probably smaller than those of today. Fossils, the ancient remains of living organisms that have turned to stone, can reveal the climatic history of a region. For example, because scientists have found fossils of a small species of hippopotamus there, they believe that the Arabian Desert, which covers most of the Arabian Peninsula to the east of North Africa, once included wetlands. In the Sahara Desert of North Africa, rock paintings done 5,000 years ago show pictures of elephants, giraffes, and herds of antelope that could no longer survive there.

Today, "desertification" (dee-zur-tih-fih-KAY-shun; desert formation), occurs continuously, primarily on the edges of existing deserts. Desertification is caused by a combination of droughts (rainless periods) and human activity such as deforestation (cutting down forests) or overgrazing of herd animals. When every blade of grass is used and rain is scarce, plants don't grow back. Without plants to hold the soil in place, the wind blows away the smaller and finer particles of soil exposing the less compacted layer of sand, leaving a barren, unprotected surface. Eventually, even groundwater disappears. In 1882, the percentage of land classified as desert was only around 9 percent. By 1950, arid (very dry) and semiarid lands combined had grown, at a rate of about 30 miles (48 kilometers) a year, to more than 23 percent. Today, because of desertification, the Sahara in North Africa, the world's largest desert, has advanced southward another 600 miles (1,000 kilometers).

KINDS OF DESERTS

Scientists measure a region's aridity by comparing the amount of precipitation (rain, sleet, or snow) to the rate of evaporation. Evaporation always exceeds precipitation. Deserts can be classified as extremely arid (less than 1 inch [2.5 centimeters] of rain per year); arid (up to 10 inches [25 centimeters]); and semiarid (as much as 20 inches [50 centimeters]) of rain per year, but so hot that moisture evaporates rapidly. Most true deserts receive less than 4 inches (10 centimeters).

Except for those at the North and South Poles, which are special cases, deserts are generally classified as hot or cold. Daytime average temperatures in hot deserts are warm during all seasons of the year, usually above 65°F (18°C). Nighttime temperatures are chilly and sometimes go below freezing. Typical hot deserts include the Sahara and the Namib Desert of Namibia. Cold deserts have hot summers and cold winters. For at least one month during the year, the mean temperature is below 45°F (7°C). Cold deserts include the Turkestan in Kazakhstan and Uzbekistan, the Gobi (GOH-bee) in China and Mongolia, and the Great Salt Lake Desert in Utah. These deserts usually get some precipitation in the form of snow.

Deserts can be further characterized by their appearance and plant life. They may be flat, mountainous, broken by gorges and ravines, or covered by a sea of sand. Plants may range from nearly invisible fungi to towering cacti and trees.

CLIMATE

Although desert climates vary, they are always extreme.

TEMPERATURE

In hot deserts, days are usually sunny and skies cloudless. During the summer, daytime air temperatures between 105° and 110°F (43.8° and

---

WELL-KNOWN DESERTS OF THE WORLD

Name	Type	Location	Approximate Size
Arabian	Hot; extremely arid and arid	Arabian Peninsula	900,000 square miles (2,330,000 square kilometers)
Atacama	Hot; extremely arid and arid	Chile	54,000 square miles (140,000 square kilometers)
Australian	Hot; arid and semiarid	Australia	600,000 square miles (1,500,000 square kilometers)
Death Valley	Hot; arid	California (United States)	5,312 square miles (13,812 square kilometers)
Gobi	Cold; arid and semiarid	China, Mongolia	500,000 square miles (1,300,000 square kilometers)
Kalahari	Hot; arid	Southern Africa	19,300 square miles (500,000 square kilometers)
Mojave	Hot; arid	California and Nevada (United States)	25,000 square miles (65,000 square kilometers)
Namib	Hot; arid	Botswana, eastern Namibia, and northern South Africa	52,000 square miles (135,000 square kilometers)
Negev	Hot; arid	Israel	4,700 square miles (12,170 square kilometers)
Patagonian	Cold; arid	Argentina	260,000 square miles (673,000 square kilometers)
Sahara	Hot; extremely arid and arid	North Africa	3,200,000 square miles (8,600,000 square kilometers)
Thar	Hot; extremely arid	India and Pakistan	77,000 square miles (200,000 square kilometers)

46.8°C) are not unusual. A record air temperature of 136.4°F (62.6°C) was measured in the Sahara Desert on September 13, 1922, and that was in the shade! Because there is little vegetation, rocks and soil are exposed to the Sun, which may cause ground temperatures in the hottest deserts to exceed 175°F (80°C). Nights, however, are much cooler. The lack of cloud cover allows heat to escape and the temperature may drop 25 degrees or more after the Sun sets. At night temperatures of 50°F (10°C) or less are common, and they may even drop below freezing.

Winters in cold deserts at latitudes midway between the polar and equatorial regions can be bitter. In the Gobi Desert, for example, temperatures below freezing are common. Blizzards and violent winds often accompany the icy temperatures.

RAINFALL

Rainfall also varies from desert to desert and from year to year. The driest deserts may receive no rainfall for several years, or as much as 17 inches

THE SANDS OF TIME

When living things die, moisture in the air aids the bacteria that cause decay. Before long, tissues dissolve and eventually disappear. Desert air is so dry that decay does not take place or occurs extremely slowly. Instead, tissues dry out and shrink, turning an animal or human being into a mummy.

In ancient Egypt around 3000 B.C., the dead were buried in shallow graves in the sand. The very dry conditions mummified the bodies, preserving them. Later, for those who could afford it, Egyptian burials became more complex. Internal organs were removed, and the bodies underwent special treatments designed to preserve them. Finally, they were placed into tombs dug into the rocky cliffs or, in the case of certain pharaohs (kings), placed within huge pyramids of stone. In most cases, bodies of the ancient Egyptians are so well preserved that much can still be learned about what the

people ate, how they lived, and what caused their deaths.

Recently, graves discovered in the Takla Makan (TAHK-lah mah-KAN) Desert of China have also given scientists important information. (The name Takla Makan means "the place from which there is no return.") Well-preserved mummies as much as 3,800 years old have been found in the graves. The mummies have European features and some are dressed in fine woolens woven in tartan (plaid) patterns that were commonly used by the ancient Celts and Saxons of Northern Europe. Scientists believe that these mummies were the first Europeans to enter China, which was officially closed to outsiders for thousands of years. Evidence exists that these people rode horses using saddles as early as 800 B.C. and that they may have introduced the wheel to China. Their descendants, who have intermarried with the Chinese, still live in the Takla Makan.

(430 millimeters) in a single year. Rainfall may be spread out over many months or fall within a few hours. In the Atacama Desert of Chile, considered the world's driest desert, more than half an inch (12.5 millimeters) of rain fell in one shower after four years of drought. Such conditions often cause flash floods, which sweep vast quantities of mud, sand, and boulders through dry washes, gullies, and dry river beds (sometimes called *wadis* or *arroyos*). The water, however, soon evaporates or disappears into the ground. The Atacama is also the site of the world's longest drought, where no rain fell for 400 years (from 1571 until 1971).

In coastal deserts, fog and mist may be common. Fog occurs when cold-water ocean currents cool the air and moisture condenses.

GEOGRAPHY OF DESERTS

The geography of deserts involves landforms, elevation, soil, mineral resources, and water resources.

LANDFORMS

Desert terrain may consist of mountains, a basin surrounded by mountains, or a high plain. Many desert areas were once lake beds that show the effect of erosion and soil deposits carried there by rivers. Wind also helps shape the desert terrain by blowing great clouds of dust and sand that bite into rock, sometimes sculpting it into strange and magnificent shapes. In the Australian Desert, unusual pinnacles (tall mountain shapes) of limestone rock formed over thousands of years by the wind, stand on the flat desert floor.

Sand dunes Bare rock, boulders, gravel, and large areas of sand appear in most desert landscapes. However, vast expanses of sand dunes, sometimes called ergs, occur less frequently than many people believe. Sand dunes make up less than 2 percent of deserts in North America, only 11 percent of the Sahara, and 30 percent of the Arabian Desert. The Empty Quarter (*Rub al Khali*) in the Arabian Desert is the largest area of dunes in the world, covering about 250,000 square miles (647,500 square kilometers).

Unless anchored by grass or other vegetation, sand dunes migrate constantly. Their rate of movement depends upon their size—smaller dunes move faster—and the speed of the wind. Some dunes may move over 100

DESERT BE GOOD

Becoming lost in the desert can often end in tragedy as the following example proves. In 1943, an American bomber, called the *Lady Be Good,* went off course and crashed in the Sahara Desert. The surviving crew saw a line of hills in the distance and mistook them for the hills around the Mediterranean, where they hoped to find human settlements. At night they walked and they rested by day, covering 75 miles (120 kilometers) in a week. But, as it turned out, the hills were not those around the Mediterranean, and the crewmen were still 375 miles (600 kilometers) from the sea. The entire crew died from exposure and lack of food and water, and their bodies were not found until 1960—17 years later.

feet (30 meters) in a year and can bury entire villages. In the Sahara, dunes created by strong winds may achieve heights of 1,000 feet (305 meters). Larger dunes may become permanent. Scientists estimate that dunes in the Namib Desert may be as much as 40,000,000 years old.

Dunes take different shapes, depending upon how they lie in respect to prevailing winds. When the wind tends to blow in one direction, dunes often form ridges. The ridges may lie parallel to the wind, forming seif (SAFE), or longitudinal dunes, or at right angles to it forming transverse dunes. Seif dunes are the largest, with some in the Sahara approaching 250 miles (400 kilometers) in length. At desert margins, where there is less sand, dunes may assume crescent shapes having pointed ends. These are called barchan (bahr-KAN) dunes, and the wind blows in the direction of their "points." In the Sahara, stellar (star-shaped) dunes are commonly found. Stellar dunes are formed when the wind shifts often, blowing from several directions.

ELEVATION

Deserts exist at many altitudes. North American deserts are partly mountainous, but Death Valley, a large basin in California, is 282 feet (86 meters) below sea level at its lowest point. The main plateau of the Gobi is 3,500 feet (1,067 meters) above sea level, and the Sahara extends from 436 feet (133 meters) below sea level to 11,204 feet (3,361 meters) above. Temperature, plant, and animal life, are all influenced by elevation.

SOIL

Desert soils tend to be coarse, light colored, and high in mineral content. They contain little organic matter because there is so little vegetation. If the area is a basin or a catch-all for flash-flood waters, mineral salts may be carried to the center where concentrations in the soil become heavy. If the area was once an inland sea, like the Kalahari (kah-lah-HAHR-ee) Desert of Botswana, eastern Namibia, and northern South Africa, exposed bottom sediments (matter deposited by water or wind) are very high in salt.

Most desert sand is made of tiny particles of the mineral quartz. Placed under pressure for long periods, grains of sand may stick together, forming a type of rock called sandstone.

Some deserts have little soil, exposing bare, wind-polished, pebbly rock, called "desert pavement." Rocks are often broken due to contraction and expansion caused by extreme temperature variations. Basin areas scoured by winds often show surfaces of gravel and boulders, and on steep slopes, whipping winds may leave little soil.

OPPOSITE:
An illustration showing how the wind forms the four different types of sand dunes in a desert.

Dust devils, columns of dust that spin over the desert landscape, are carried by whirlwinds. Some dust storms produce clouds thousands of feet high. In the Sahara, up to 200,000,000 tons (180,000,000 metric tons) of

Types of Sand Dunes
(Arrows indicate wind direction)

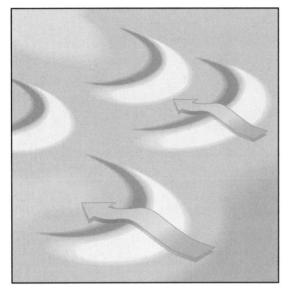

Barchan Dunes

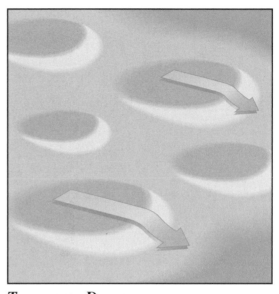

Transverse Dunes

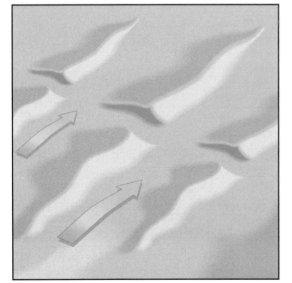

Seif Dunes

Stellar Dunes

dust is created each year. Red Saharan dust has been found on rooftops as far away as Paris, France, more than 1,000 miles (1,600 kilometers) from the Sahara.

Two long-lasting chemical reactions also affect desert rocks and soil. One, called desert varnish, gives rocks, sand, and gravel a dark sheen. Desert varnish is believed to be caused by the reaction between the moisture from overnight dew and minerals in the soil. The second reaction is the formation of duricrusts—hard, rocklike crusts that form on ridges when dew and minerals such as limestone combine, creating a type of cement.

Desert soils offer little help to plant life because they lack the nutrients provided by decaying vegetation and are easily blown away, exposing plant roots to the dry air. Some deep-rooted plants can exist on rock, however, where moisture accumulates in cracks. Other plants remain dormant during the driest periods, thriving and blooming after brief rains.

Soil also reveals much about a desert's geological history. In Jordan, a Middle-Eastern country, for example, the Black Desert takes its name from black basalt, a rock formed from volcanic lava.

MINERAL RESOURCES

Valuable minerals like gold and oil (petroleum) are often found in desert regions. In the Great Sandy Desert of Australia, miners hunt for gold nuggets. "Black gold," as oil is often called, is found beneath the desert regions of the Middle East, where it was formed over time from the sediment of prehistoric oceans. (Countries such as Saudi Arabia have become wealthy from the sale of their oil reserves.) Iron ore is mined in portions of the Sahara. And, borax—a white salt used in the manufacture of such products as glass and detergent—was once mined in Death Valley, California.

WATER RESOURCES

Water sources in the desert include underground reserves and surface water.

Groundwater In addition to occasional rainfall, deserts may have reserves of underground water. These reserves, often trapped in layers of porous rock called aquifers, were formed over thousands of years when rainwater seeped underground. Reserves close to the surface may create an oasis, a green, fertile haven where trees and plants thrive. The presence of water may allow a completely different biome to form like an island in the desert.

Desert peoples often dig wells into aquifers and other underground water sources to irrigate crops and water their animals. As desert populations grow, however, water sources shrink and cannot be replaced fast enough. There is a real danger that groundwater reserves will one day be depleted.

Surface water Water may also be found in desert areas in the form of rivers or streams. Some streams form only after a rain, when water sweeps along a dry river bed in a torrent (violent stream) then quickly sinks into the ground or evaporates. However, moisture sometimes remains under the surface, for plants can be seen growing along the path of streams.

Permanent rivers are also found in desert regions. The Colorado River is one example. Over a million years ago, the Colorado began to cut a path into the plateau of limestone and sandstone rock in northern Arizona, ultimately forming the Grand Canyon, which is 1.2 miles (1.9 kilometers) deep and 277 miles (446 kilometers) long. Perhaps the most famous desert river is the Nile, which bisects Egypt. Since ancient times, Nile floods have brought enough rich soil from countries farther south to turn Egypt's river valley into fertile country well known for its agricultural products, such as cotton.

Permanent lakes, however, rarely occur in desert regions. Two exceptions are the Great Salt Lake of Utah, which is all that remains of what once was a great inland sea, and the Dead Sea of Israel and Jordan. The Dead Sea is actually a salt lake that was once part of the Mediterranean.

PLANT LIFE

One of the most important characteristics of any biome is its plant life. Not only do plants provide food and shelter for animals, they recycle gases in the atmosphere and add beauty and color to the landscape. Deserts support many types of plants, although not in large numbers.

ALGAE, FUNGI, AND LICHENS

Although it is generally agreed that algae (AL-jee), fungi (FUHN-jee), and lichens (LY-kens) do not fit neatly into either the plant or animal categories. In this chapter, however, these special organisms will be discussed as if they, too, were plants.

Algae Most algae are single-celled organisms; a few are multicellular. Certain types of algae make their own food by means of photosynthesis, while others absorb nutrients from their surroundings. (Photosynthesis is the process by which plants use the energy from sunlight to change water and the air's carbon dioxide into the sugars and starches they use for food.) Although most algae are water plants, blue-green algae do appear in the desert. They survive as spores during the long dry periods and return to life as soon as it rains. (Spores are single cells that have the ability to grow into a new organism.)

Fungi Fungi are commonly found in desert regions wherever other living organisms are found. They, too, reproduce by means of spores. Fungi cannot make their own food by means of photosynthesis. Some, like mold and mushrooms, obtain nutrients from dead or decaying organic matter. They assist in the decomposition (breaking down) of this matter, releasing nutrients needed by other desert plants. Other fungi are parasites and attach themselves to other living plants. Parasites can be found wherever green plants live, and some often weaken the host plant so that it eventually dies. Others actually help their host absorb nutrients more effectively from the soil.

Lichens Lichens are actually combinations of algae and fungi living in cooperation. The fungi surround the algal cells. The algae obtain food for themselves and the fungi by means of photosynthesis. It is not known if the fungi aid the algal organisms, although they may provide them with protection and moisture.

Lichens are among the plants that live the longest. Some living in polar deserts are believed to have survived at least 4,000 years. Although lichens grow slowly, they are very hardy and can live in barren places under extreme conditions, such as on bare desert rock or arctic ice. Crusty types, colored grey, green, or orange, often cover desert rocks and soil. During very dry periods, they rest. When it rains, they grow and make food.

GREEN PLANTS

Most green plants need several basic things to grow: light, air, water, warmth, and nutrients. In the desert, light, air, and warmth are abundant, although water is always scarce. The remaining nutrients—primarily nitrogen, phosphorus, and potassium—which are obtained from the soil, may be in short supply.

Because water is limited, desert plants must protect against water loss and wilting, which can damage their cells. Large plants require strong fibers or thick, woody cell walls to help hold them upright. Even smaller plants have a great number of these, which makes them fibrous and tough. Their leaves tend to be small and thick, with less surface exposed to the air. Outer leaf surfaces are often waxy, which helps prevent water loss. Pores in the surface of green leaves allow the plant to "breathe," taking in carbon dioxide and releasing oxygen. The leaves of some desert plants may have grooves to protect their pores against the movement of hot, dry air. Other leaves curl up or develop a thick covering of tiny hairs for protection. Still others have adjusted to the dry environment by adapting the shape of their leaves. For, example cactus leaves are actually spines (needles). These spines provide less surface area from which water can evaporate. As a result, more water is stored within the plant.

Growing season The growing season in deserts is limited to the brief periods of rain, which, in some cases, do not occur for several years. In some coastal deserts, certain plants absorb mist from the nearby ocean through their leaves. Where the soil is rich and rain more regular and dependable, desert plants may flourish.

In cold deserts, where real winter occurs, plants behave much like those in temperate climates. The portion above ground dies off, but the root system goes deep and is protected from freezing by layers of snow.

Deserts are home to both annuals and perennials. Annuals live only one year or one season and they require at least a brief rainy period that occurs regularly. Their seeds seem to sprout and grow overnight into a sea of colorful, blossoming plants. This period of rapid growth may last only a few weeks. When the rains disappear, the plants die and the species withdraws into seed form, remaining dormant (asleep) until the next period of rain.

Unlike annuals, perennials live at least two years or two growing seasons, appearing to die in between but returning to "life" when conditions improve. Those that live many years must be strong and use several methods to survive. During their youth, these plants devote most of their energy to developing a large root system to collect any available moisture. Plants are often many yards apart, because their roots require a large area of ground in order to find enough water. The above-ground portion of young perennials is small in comparison to the root system because their leaves do not have to compete for sunlight or air as they do for water. Some perennials are succulents (SUHK-yoo-lents), which are able to store water during long dry periods. Some, like the century plant, store water in their leaves, while others store it in their stems or in large roots.

THE WEB OF LIFE: EARTH AS A SYSTEM

Many scientists now recognize the many links among all life forms living on our planet. One biologist, James Lovelock, has theorized that life itself is responsible for changes in the land, water, and air. For example, until about 2,000,000,000 years ago, there was almost no oxygen in the atmosphere. Then plants began using energy from the Sun for photosynthesis (a food-making process), which produces oxygen as a by-product. After enough plants got to work, the atmosphere eventually became, and is still maintained at, 21 percent oxygen, which is ideal for animal life. Lovelock believes that living things somehow work together in this way, instinctively providing a comfortable environment for themselves and one another.

To prove his theory, Lovelock has produced a computer model. (Computer models enable scientists to more quickly study processes that take very long periods of actual time to show a result.) Suppose, for example, that there are two species of flowers, one white and one dark blue. The white flowers reflect the Sun's heat and can survive in warm climates. However, the dark blue flowers absorb heat and do better in cooler climates. According to Lovelock's model, the flowers help control the environment. The presence of blue flowers in a cool environment means heat is absorbed and the surrounding temperature is prevented from being too cool. The white flowers, on the other hand, reflect the heat and help keep a warm environment from being too warm.

Water supply Except for occasional rivers and oases, water in the desert comes from the brief rains. Plants may grow in greater numbers in arroyos or wadis where some moisture may remain beneath the surface. In coastal deserts, some plants absorb moisture from fog and mists that condense on their leaves, while in cold deserts, spring thaws provide water from melting snow.

Reproduction Pollination (the transfer of pollen from the male reproductive organs to the female reproductive organs of plants) is also often a problem in deserts. Although the wind may carry pollen from one plant to another, this method is not efficient because plants usually grow far apart. Also, insect pollination is rare because there are fewer insects than in other biomes. As a result, most plants have both male and female reproductive organs and pollinate themselves.

Common desert plants Several species of plants grow in the desert, including cacti, shrubs, trees, palms, annuals.

CACTI Cactus plants originated in southern North America, Central America, and northern South America. Instead of leaves, a cactus has spines, which come in many forms from long, sharp spikes to soft hairs. Photosynthesis takes place in the stems and trunk of the plant. Nectar, a sweet liquid that appeals to insects, birds, and bats, is produced in the often spectacular flowers.

The largest plant in the desert is the giant saguaro (sah-GWAH-roh) cactus. Its large central trunk can grow as tall as 50 feet (15 meters) and weigh 11 tons (10 metric tons). Ninety percent of its weight is in water, which it stores in its soft, spongy interior. During very dry conditions, as the plant uses up this stored water, the trunk shrinks in size.

Like many desert plants, the saguaro has a wide, shallow root system designed to cover a large area. After a long dry period, its roots can take up as much as 1.1 tons (1 metric ton) of water in 24 hours. The trunk then expands as it absorbs and stores the new supply.

Pores run in deep grooves along the saguaro's stems. These pores open during the cooler nighttime hours to take in carbon dioxide and release oxygen. For protection from wind and animals, long, sharp spines run along the grooves. These spines reduce air movement which conserves water and keeps grazing animals away.

Birds and bats love the nectar produced in saguaro flowers, and bats help to pollinate the saguaro. A bat's head fits the shape of the flowers almost perfectly and, as the bat drinks the nectar, its head gets heavily dusted with pollen grains, which it then carries to the next plant.

> **A PRICKLY COMPASS**
>
> The stems of the barrel cactus of the southwestern United States grow in a curve, and that curve always points south. This happens because the cactus grows faster on its northern side, which is in the shade.

WOODY SHRUBS AND TREES Small, woody perennials that flourish in the desert climate include sagebrush, salt-brush, creosote, and mesquite. They have small leaves and wide-ranging root systems. Most shrubs have spines or thorns that protect them against grazing animals.

The mesquite (meh-SKEET) is a tree with roots that may grow 100 feet (30 meters) deep. Because of its long roots, it manages to find a constant supply of water and remains green all year. It produces long seed pods with hard, waterproof coverings. When the pods are eaten by desert animals, the partially digested seeds pass out of the animal's body and begin to grow.

Joshua trees, a form of yucca plant, live for hundreds of years in the Mojave Desert of California. Yuccas and treelike aloes store water in their leaves, not their stems. One type of aloe, the kokerboom tree of southwest Africa, can survive several years without rain.

PALMS Date palms, found at many oases in the Sahara and Arabian Deserts, can grow in soil with a high salt content. Only female trees produce dates, and only a few male trees are necessary for cross-pollination. The dates can be eaten raw, dried, or cooked and are an important food source for desert dwellers.

The Washingtonian fan palm is the largest palm in North America. Native to California, it requires a dependable supply of water to survive and sends out a thick web of tiny roots at its base. When its fronds (branches)

Saguaro cacti among volcanic peaks in the Sonoran Desert of Arizona and Utah. (Reproduced by permission of Corbis. Photograph by David Muench.)

die, they droop down around the trunk and form a "skirt," where animals like to live. The fan palm produces a datelike fruit, which is eaten by many desert inhabitants.

ANNUALS Long grasses, such as alpha or esparto grass, often flourish in the desert after seasonal rains. Their stems can be used to make ropes, baskets, mats, and paper. Tufts of grasses sometimes become so entwined that they form balls which, when blown by the wind, drop their seeds as they spin across the landscape.

Almost every desert has its share of blooming annuals that add masses of color after a rain. Primroses and daisies adorn the California desert, while daisies, blue bindweed, dandelions, and red vetch beautify the Sahara.

ENDANGERED SPECIES

Many desert plants, such as the candy cactus, saguaro cactus, and the silver dollar cactus, are nearing extinction because of their popularity as

A Joshua tree, a type of yucca plant, blooming in the Mojave Desert of California. (Reproduced by permission of National Audubon Society Collection/Photo Researchers, Inc. Photograph by Verna Johnston.)

house plants and for landscaping. Even though they are available for sale in nurseries, some people steal them from the wild. Because desert plants are sparse to begin with, removal from their native home upsets the delicate balance of their reproduction.

ANIMAL LIFE

All animals face the same problems in adapting to the desert. They must find shelter from daytime heat and nighttime cold, as well as find food and water, which are often scarce. Yet, in spite of these extreme conditions, most animal species are represented in the desert environment, even some we typically associate with temperate or wet surroundings.

INVERTEBRATES

Animals without backbones are called invertebrates. They include simple desert animals such as worms, and more complex animals such as the locust. Certain groups of invertebrates must spend part of their lives in water. Generally speaking, these types are not found in deserts. One exception is the brine shrimp, an ancient species that can live in desert salt lakes. Other exceptions are certain species of worms, leeches, midges, and flies that live in the fresh water of oases and other waterholes.

A cluster of date palms at the Nefta Oasis in the Sahara Desert. (Reproduced by permission of Corbis. Photograph by Wolfgang Kaehler.)

Most invertebrates are better adapted to desert life than vertebrates. Many have an exoskeleton (an external skeleton, or hard shell, made from a chemical substance called chitin [KY-tin]). Chitin is like armor and is usually waterproof. It also protects against the heat of the desert Sun, preventing its owner from drying out.

Food and water Many invertebrates are winged and can fly considerable distances in search of food. Many eat plant foods or decaying animal matter. Some invertebrates are parasites, like the Guinea worm, which lurk at water-holes waiting for some unsuspecting animal to wander by. Parasites attach themselves to the animal's body or are swallowed and invade the animal from the inside.

The arachnids (spiders and scorpions), which are carnivores (meat eaters), seem well suited to desert life. They prey on insects and sometimes, if they are large enough, small lizards, mice, and birds. Scorpions use their pinchers to catch prey, then inject it with venom (poison) from the stinger in their tails. Their venom is dangerous, even for large animals and humans, and can kill. Arachnids do not usually drink water but get what they need from their prey.

Shelter Most desert spiders live on the ground rather than on webs, hiding in holes or under stones to escape the heat. The large, hairy, camel spiders, which live in all deserts except those of Australia, are nocturnal, which means they rest during the daytime hours and hunt at night.

Scorpions, too, hunt at night and avoid the heat of the day, taking shelter beneath rocks or burying themselves in sand or loose gravel. Several American and Australian species dig burrows that may reach more than 3 feet (1 meter) below the surface.

Reproduction Some invertebrates, such as the scorpion, go through a mating ritual. Male and female scorpions appear to "dance," their pinchers clasped together, as they do a two-step back and forth. After mating, the males hurry away to avoid being killed and eaten by the female.

Most invertebrates have a four-part life cycle that increases their ability to survive in a hostile (unfriendly) environment. The first stage of this cycle is the egg. The egg's shell is usually tough and resistant to long dry spells. After a rain and during a period of plant growth, the egg hatches. The second stage is the larva (immature), which may actually be divided into several stages between which there is a shedding of the outer covering, or skin, as the larva increases in size. Larvae have it the easiest of all in the desert, often being able to spend a portion of their life cycle below the ground where it is cooler and more moist than on the surface. Some larvae store fat in their bodies and do not even have to seek food. The third stage of development is the pupal stage. During this stage, the animal often lives inside a casing, in a

A woman standing next to a termite mound, showing how tall they may become. (Reproduced by permission of Edward S. Ross.)

resting state, which may offer as much protection as an egg. Finally, the adult emerges.

Common desert invertebrates Termites and locusts are both invertebrates found in the desert.

TERMITES Termites, found all over the world, build the skyscrapers of the desert. Their mounds, often more than 6 feet (2 meters) tall, are erected over a vast underground system of tunnels. These tunnels usually go as deep as the groundwater so that a water supply is readily available to the termite colony. The mounds, which are made of dirt, decaying plants, and termite secretions that dry rock-hard in the Sun, have many air ducts. As the Sun warms the mounds, they grow very hot. As the hot air inside the mounds rises, cooler air is drawn upward through the tunnels, creating a type of air-conditioning system.

Termites eat plant foods, especially the cellulose (substance making up the plant's cell walls) found in woody plants.

Termites have an elaborate social structure. A single female—the queen—lays all the eggs and is tended to by workers. Soldier termites, equipped with huge jaws, guard the entrances to the mound. Soldiers cannot feed themselves, and the workers must tend to them, as well.

LOCUSTS Locusts are found in the deserts of northern Africa, the Middle East, India, and Pakistan. Similar to grasshoppers in appearance, they have wings and can fly, as well as leap, for considerable distances. For years they live quietly, nibbling on plants and producing a modest number of young. Then, for reasons not completely understood, their numbers increase dramatically. Suddenly, great armies of locusts emerge, hopping or flying through the desert in search of food. Eating every plant in sight, these swarms may travel thousands of miles before their feeding frenzy ends. In a short time, the hordes die off and locust life returns to normal. However, the devastated landscape they leave behind may take years to recover.

> ## LIVING JUICE BOXES
>
> Honeypot ants, which live in Africa, Australia, and the Americas, take the business of food very seriously. They maintain "herds" of aphids, insects that suck the juices of plants and then secrete a sugar-rich "honey." Worker ants collect this honey and bring it back to the nest where they feed it to a second group of workers. This second group takes in so much aphid honey that their stomachs swell and it is difficult for them to move. They then suspend themselves from the ceiling of the nest and wait. Later, when food becomes scarce, they spit the honey back up for other ants to eat. Some desert peoples like eating honeypot ants and seek out their nests.

AMPHIBIANS

Amphibians are vertebrates (animals with backbones) that usually spend part, if not most, of their lives in water. Unlikely as it seems, such animals can be found in a desert. Frogs and toads both manage to survive in significant numbers in desert environments.

The short, active portion of their lives occurs during and immediately after the seasonal rains, when pools of water form. Mating, egg-laying, and young adulthood all take place in these pools. Those that survive into maturity leave the pools and take their chances on the desert floor where they are able to spend a few weeks feeding on both plants and insects. They must find shade, however, or risk dying in the heat of the Sun.

Food Frogs and toads feed on algae, plants, and freshwater crustaceans such as tiny shrimps that manage to survive in egg or spore form until brought to life by the rains. Frogs that eat the meat of crustaceans while they are tadpoles often become cannibals as they mature, eating their smaller, algae-eating brothers and sisters. If the rainy season is short, the cannibals have a better chance of survival because they have more food choices. If the rainy season lingers, however, the smaller tadpoles have a better chance because the cannibals can't see their prey as well in muddy waters churned up by the rains, but the plant-eaters receive an increased supply of algae.

Shelter During the hottest, driest seasons, amphibians go through estivation (ess-tih-VAY-shun), an inactive period. While the soil is still moist from the rain, they dig themselves a foot or more into the ground. Only their nostrils remain open to the surface. Normally, their skin is moist and soft and helps them absorb oxygen. During estivation, however, the skin hardens and forms a watertight casing. All the animal's bodily processes slow down to a minimum, and it remains in this state until the next rainfall, when it emerges. When water is scarce, Australian Aborigines (native peoples) dig up estivating frogs or toads and squeeze the animal's moisture into their mouths.

> ### AVOIDING A HOT FOOT
> Several species of lizards have evolved methods for traveling over hot sand without burning their toes. The agamid lizard has long legs and keeps one foot raised and swinging in the breeze to cool it off while the other three support the animal. It does this in rotation, so that all its feet get an equal chance to cool off.

Reproduction Mating and egg-laying for amphibians must take place in water, because male sperm are deposited in the water and must be able to swim to the jelly-like eggs in order to penetrate them. As the young develop into larvae and young adults, they often have gills. Therefore, they too require a watery habitat. If there is not enough rain for pools of water to form, amphibian populations may not survive.

REPTILES

Of all the animals, reptiles are perhaps most suited to living in the desert. Those most commonly found there are snakes, lizards, and some species of tortoises. Their scaly, hard skin prevents water loss, and their urine is almost solid, so no water is wasted.

Reptiles are cold-blooded, which means that their body temperature changes with the temperature of the surrounding air. This may actually help them live in the desert. Early in the day, they expose as much of their bodies as possible to the Sun for warmth. As the temperature climbs, they expose less and less of their bodies. During the hottest period of the day, they find shade or a hole in which to wait for cooler temperatures. During the chilly nights, although they become sluggish, they do not have to waste energy keeping their body temperatures up as most mammals and birds must.

> ### THE LIVING FUNNEL
>
> Most lizards obtain water from the food they eat. The thorny devil, a small lizard from Australia, is an exception, however. During the cool nights, dew condenses in the creases and folds of its skin. Most of these folds lead toward the lizard's mouth, and the lizard is able to lap up the moisture.

Besides being cold-blooded, some species of lizards have specially developed clear membranes in their lower eyelids that cover the entire eye and protect the eye from lost moisture.

Snakes have no legs. They move using special muscles that flex their flat belly scales forward and backward. Ridges on their scales grip the ground and pull them along. Some rattlesnakes, like the sidewinders of North American deserts, manage to move diagonally by coiling into a kind of S-shape, and propel themselves by pushing with the outside back portion of each s-shaped curve.

Food and water The diet of lizards varies, depending upon the species. Some have long tongues with sticky tips that are good for catching insects. Many are carnivores that eat small mammals and birds. The water they need is usually obtained from the food they eat.

While most lizards hunt food during the day, the gila monster, found in the southwestern United States, looks for reptile eggs and baby animals after dark. Making the most of its opportunities, the lizard stuffs itself. During periods when food is scarce, its body draws on extra nutrients stored as fat in its tail, which can double in size after a big meal.

All snakes are carnivores, and one decent-sized meal will last them for days or weeks. In the desert, they make good use of their eyes to hunt during the cool nights when their prey are most active. Snakes cannot close their eyes because they have no eyelids. However, a transparent covering protects their eyes from the dry air, dust, and sand. Because they often hunt underground they have adapted to detecting ground vibrations. Many desert snakes bury themselves in the sand so that only their eyes and flickering tongues are visible. There they wait. The more evolved snakes kill their prey with venom (poison).

Although they are commonly thought of as jungle dwellers, boa constrictors and pythons (also a species of constrictor) also live in the desert. Constrictors strike their prey, hold it with a mouthful of tiny teeth, then wrap their body around it like a coil. Gradually, the prey suffocates and the constrictor swallows its meal whole, gradually working it into its stomach with its hinged lower jaw and strong throat muscles.

Shelter Lizards and snakes retire to the shade during the hottest hours of the day, where they can escape the Sun. Only a few make their own burrows. Most take over the abandoned burrows of other animals, find shelter in rock crevices, or bury themselves in the sand.

Desert tortoises obtain some shade from the Sun with their thick outer shells. But, most of the time they escape the heat of the day by retreating to burrows, which they have dug. In the spring and autumn, when the days are not excessively warm, the tortoise ventures out during the day to forage for food. However, during the summer, they forage for food at night when it is cooler, remaining in burrows during the day. In the winter, tortoises hibernate (become dormant) in a second burrow, which they have dug.

> ### THE SECRET WEAPON OF PIT VIPERS
> Rattlesnakes and other pit vipers have small pits on both sides of their faces. These pits can detect the body heat of prey, just as heat-seeking missiles do. Pit vipers hunt at night, and their secret weapon allows them to find small animals hidden in the dark.

Reproduction The eggs of reptiles are leathery and tough and do not dry out easily. Some females remain with the eggs, but most bury them in a hole. Offspring are seldom coddled and are left to hatch by themselves. Once free of the eggs, the babies dig themselves out of the hole and begin life on their own.

Common desert reptiles Snakes are less common in deserts than lizards. Common desert snakes include the gopher snake, horned viper, Gaboon viper, rattlesnake, and cobra. The cobra is found in Africa and India.

Common lizards include the gecko—which can survive long periods without food by living on stored fat—the skink, the bearded dragon, the iguanid lizard, and the monitor lizard.

BIRDS

All deserts have bird populations. Tropical and subtropical deserts are visited twice each year by hundreds of species of migratory birds traveling from one seasonal breeding place to another. These migrators include small birds such as wheatears, as well as larger species such as storks and cranes. Some species know the route and where to find food or water. Others fly at night, when it is cool. However, migrators are not true desert birds. They cannot survive for long periods in the desert as can birds for which the desert is home.

Because birds have the highest body temperature of any animal—around 104°F (40°C)—they do not need to lose body heat until the desert temperature is greater than their own. This makes desert life easier for them than for mammals, which must lose heat regularly during the warmest months, usually by panting or sweating.

Feathers protect birds not only from the cold in winter but from the Sun and heat. Air trapped between layers of feathers acts as insulation. Birds do not sweat but, by flexing certain muscles, can make their feathers stand erect. This allows them to direct cooling breezes to their skin. Those having broad wing spans, such as eagles and buzzards, can soar at high altitudes and find cooler temperatures.

Food and water Birds are found in greater variety and numbers around oases and waterholes where there is an ample supply of water, seeds, and insects. Some, like the Australian scarlet robin, often drink water. Birds that live in the desert itself are able to fly long distances in search of food or water. Some birds become nomads, following the rains from habitat to habitat. However, birds usually require less than 10 percent of the amount of water needed by mammals. For this reason, many, like shrikes and some wheatears, can obtain enough moisture from the seeds, plants, and insects that they eat and do not need an additional water source. The same is true of vultures and birds of prey, which obtain water from the flesh of animals. Also, birds' kidneys are very efficient in their ability to extract water, and their urine is not liquid but jelly-like.

Shelter During the hottest part of the day, most desert birds rest by roosting in the shade or in underground burrows. During excessively hot or dry periods, birds can simply fly to more comfortable regions. It has been estimated that one-third of Australian birds are constantly on the move to escape the heat.

> ## THE FLYING SPONGE
>
> The sandgrouse, found in the deserts of Africa and Asia, is a water-drinking bird. Most members of this species live near waterholes, except during nesting periods when they may be forced to remain in an area suffering from drought (extreme dryness). Male sandgrouse have evolved an unusual method of overcoming this problem, however. They fly to the nearest waterhole—perhaps 20 miles (35 kilometers) away—and turn themselves into sponges. They wade into the water and let their special belly feathers absorb the liquid—as much as twenty times their own weight. Then they fly back to the nest where the nestlings drink by squeezing the water out with their beaks.

Because the desert is home to so few trees, many desert birds build nests in rock crevices, in abandoned burrows, or on the ground in the open. Those that build on the ground may put walls of pebbles around the nest which act as insulation and reduce the force of the wind.

Reproduction Except in Australia, desert birds appear to breed as other birds do—according to the seasons. In Australia, they adapt their breeding habits to periods of rainfall, and breeding cycles may be years apart.

Although birds are free to fly away from the heat during the rest of the year, during the breeding cycle they must remain in the same spot from the time nest building begins until the young birds can fly. This is usually a period of many weeks.

Normally, the parents sit on the nest to protect the eggs from heat or cold. During very hot weather, the parents may stand over the nest to give the eggs or the nestlings shade.

Common desert birds Common desert birds include ground birds and birds of prey.

GROUND BIRDS "Ground" birds are not hunters or scavengers (animals that will eat decaying matter) but obtain most of their food from plants and insects. They have strong legs that enable them to dart around on the ground without tiring.

In Asia, Africa, Australia, and the Arabian Peninsula, families of thrushes called chats are common. Varieties of chats live at many different altitudes, including those over 13,000 feet (4,000 meters). They are found in both arid and semiarid regions, and their diets and habits vary according to their location.

Wrens are also common in desert habitats all over the world. Wrens eat insects, although those in North American deserts also eat seeds and soft fruits. Cactus wrens, as their name implies, live among prickly desert plants where they build their nests among the spines.

A cobra, one of the many reptiles found in the desert, ready to attack. (Reproduced by the National Audubon Society Collection/ Photo Researchers, Inc. Photograph by Tom McHugh.)

A small desert bird that is popular as a pet is the parakeet, or budgerigar. Originally from Australia, parakeets normally live in huge flocks containing tens of thousands of birds. During years when food is plentiful, a flock may number in the millions. Parakeets prefer seeds and are nomads, often traveling from habitat to habitat in search of the seeds of annuals that bloom after a rainfall.

A desert bird that became famous as a cartoon character is the roadrunner, found in the southwestern United States. Their name is well earned, for roadrunners can scurry over the desert floor for long stretches at speeds of about 13 miles (20 kilometers) per hour. Roadrunners are carnivores and are known to come running at the sound of a creature in trouble.

> ## THE WEB OF LIFE: COMMENSALISM
> Sometimes animals are involved in relationships in which one animal is helped and the other is not harmed. This is called commensalism. One example is a desert snake that takes over the abandoned burrow of a prairie dog.

The largest ground birds found in deserts are members of the bustard family. The houbara bustard is found in the Sahara and the deserts of central Asia. It is about 2 feet (60 centimeters) tall and weighs as much as 7 pounds (3 kilograms). Although houbaras depend primarily upon plants for food, they also eat invertebrates and small lizards. Houbaras can run fast—up to 25 miles (40 kilometers) per hour—and seldom fly.

BIRDS OF PREY Birds of prey are hunters and meat eaters. They soar high in the air on the lookout for small animals for food. Their eyesight and hearing are usually very sensitive and enable them to see and hear creatures scurrying on the ground far below.

Several species of falcons live in semiarid regions, but the true desert falcon is the prairie falcon of North America. Prairie falcons hunt for other birds such as larks and quail, and small mammals such as rabbits and prairie dogs. When food is scarce, they will also eat insects and reptiles. A falcon attacks its prey by diving at the head and trying to seize the head in its talons. Prairie falcons do not build their own nests. Instead, they move into the abandoned nests of other birds or use a hollow in the rock.

> ## LOOK INTO MY EYES!
> The eyes of all birds are large in relation to their body size. Some of the largest are the eyes of owls, which take up about one-third of the space in the skull. Owls' eyes are specially adapted to see better at night. The eyes have more rods, the cells that are most sensitive to low light levels, and fewer cones, the cells sensitive to bright light. An owl's eyes are pear shaped and face forward, which gives the owl a kind of binocular vision and superior judgment of distances. However, because of their size and shape, an owl's eyes cannot turn in the socket. To look around, the owl must swivel its entire head.

Owls live on the edges of deserts. The largest, the eagle owl, measures 27 inches (68 centimeters) in length with a wingspan of 66 inches (168 centimeters) and has the power to attack small deer. The smallest owl, the elf owl, hunts invertebrates and small mammals such as mice and gerbils. This owl is about 5 to 6 inches (12 to 15 centimeters) long and weighs 1.25 to 1.75 ounces (35 to 55 grams). Most owls hunt in the evening or at night when their extraordinary eyes and superior hearing allow them to find their prey.

The bird most often pictured in the desert is the vulture, usually waiting patiently while some poor creature dies of thirst. However, most of the world's vultures are not really desert birds, although they do spend much time hunting in desert regions. They soar high in the air, circling and looking for carrion (dead animals), which their excellent eyesight allows them to see easily. When one bird sees a potential food source it begins to descend, and the other vultures follow.

MAMMALS

Many mammals live in the desert. More than 70 species live in North American deserts alone. Among all of there species, only monkeys and apes are rarely seen.

Mammals, too, must prevent the loss of moisture from their bodies in the desert. The urine and feces of many desert mammals are concentrated, containing only a small amount of water. Also, small desert mammals, such as rodents, do not sweat. They usually manage to lose enough heat through their skins, much as a radiator does. During the hottest times, they burrow underground. Some estivate (remain dormant) during summer months.

Medium-sized mammals, such as rabbits and hares, do not burrow or estivate; although they will use another animal's burrow to escape from immediate danger. They also have no sweat glands and cannot keep cool by sweating, although some heat escapes from their large ears. Hares and rabbits stay in whatever shade they can find during the hottest time of day. The quokka, a rabbit-like marsupial (a mammal that carries new offspring in a pouch) from Australia, copes with the heat by producing large amounts of saliva and licking itself. It is not known how the quokka makes up for this water loss.

Larger grazing animals, such as gazelles, can sweat, which helps them tolerate the heat. Some carnivores, such as coyotes, release body heat by panting (breathing rapidly through their mouths). However, this results in lost moisture. True desert dwellers, like the mongoose, the meerkat, and the hyena, avoid the midday heat by retiring to underground dens.

Food and water Some small mammals eat plant foods and insects; others, like the desert hedgehog, eat bird and reptile eggs and young. In Australia, a tiny mole no more than 8 inches (20 centimeters) long eats at least its own weight in insects and young lizards every day. Many small mammals do not need to drink water as often because they obtain moisture from the food they eat.

Large grazing animals need to drink and require a water source to replace the moisture lost in sweating. A few, however, such as the Arabian gazelle and the Nubian ibex, seldom drink water. They, too, obtain what they need from the plants they eat. Because they are mobile, grazing animals can travel to areas where rain has recently fallen. In temperate climates, most grazing animals live in large herds. The desert food supply, however, will not support such numbers, and groups are usually very small.

Carnivores such as mountain lions and coyotes, do not live in the deep desert but remain on the desert fringes where a supply of water can be more readily found. Although carnivores obtain much of their moisture requirements from the flesh of their prey, they still need to drink.

Shelter Small mammals remain in burrows during the day. The air inside a burrow is up to five times more humid than the outside air, and this helps the animal prevent moisture loss.

Medium-sized hares and rabbits do not live in burrows but seek shelter in the shade of plants or rocks and in shallow depressions. Grazing animals, too, look for shade during the midday heat. Young animals may lie on the ground in the shade created by the adults.

Carnivores dig their own dens, which may consist of many tunnels to accommodate a family clan. The males mark their territories with scent, usually that of urine.

MAKING DO WITHOUT SUNSCREEN

In the Sahara Desert lives the naked mole rat, sometimes called the sand puppy, an animal that seems poorly designed for desert life. It has no fur, just its wrinkly skin for protection, and it is almost blind. To make matters worse, unlike most mammals, it cannot maintain a steady body temperature. However, its name gives a clue as to how it survives. It burrows. Naked mole rats are tunneling experts, and a clan of rats will create an entire apartment complex of nurseries, storage chambers, and bedrooms. They dig so fast with their sharp buck teeth that most predators can't catch them and get just a face full of dirt for their trouble. The clan leader is always a female, and she is the only one who breeds. Other females do not mature sexually until the queen dies or they decide to leave the den to create their own clan elsewhere.

Reproduction Young mammals develop inside the mother's body where they are protected from heat, cold, and predators, although the extra weight can make it difficult for the pregnant females themselves to escape danger. Female mammals produce milk to feed their young which, in the desert, presents a problem in lost moisture. Those that live in dens must remain nearby until the young can survive on their own, which can make survival for both mother and young during drought conditions more difficult.

Common desert animals Mammals commonly found in deserts include the kangaroo, the camel, and the hyena.

KANGAROO Probably the most well-known grazing animal of Australia is the red kangaroo. Although the kangaroo will hop for long distances in search of food, water is less of a problem as they get much of the moisture they need from grasses. Also, their kidneys are efficient and produce a concentrated urine.

The kangaroo's strong legs allow them to move as fast as 20 miles (30 kilometers) an hour. Each hop can carry them as far as 25 feet (8 meters). Kangaroos are marsupials, which means that females carry their young in a pouch on their abdomen. Females with young in their pouches are seldom able to travel fast. If the female is in danger and the baby is too heavy, the mother will dump the baby out in order to escape. This may seem cruel, but if she is caught the baby will die anyway. If she escapes, she will breed again.

CAMEL Of all animals, camels are the ones most people associate with the desert. The Bactrian camel has two humps and lives in the deserts of central Asia. The one-humped Arabian camel, called a dromedary, lives in the Arabian and Sahara Deserts. Camels are well adapted for desert life. They can travel almost 100 miles (160 kilometers) in a day, for as long as four days, without drinking. When they do drink, they can take in more than 20 gallons (100 liters) in just a few minutes.

At one time it was believed that camels stored water in their humps. This, however, is not true. The hump contains fat, which is used up during long journeys and which, in lean times, may shrink. Camels also have strong teeth and the membranes that line the insides of their mouths are tough, allowing them to eat almost anything that grows in the desert, even the thorniest plants. They can drink bitter, salty water that other animals can not tolerate.

Their fur is thick and, during the hottest seasons, molts (falls out) and is replaced by new, thinner hair. Camels can sweat to reduce body temperature, but their body temperature is not constant and varies depending upon surrounding air temperatures. Their eyes have long protective lashes, and their nostrils can be closed to keep out blowing sand. Their padded, two-toed feet are well insulated against the hot desert floor.

Camels can no longer be considered wild animals, and it is doubtful that any wild groups still exist. They have been thoroughly domesticated

THE RABBIT PLAGUE

In the nineteenth century, English settlers introduced rabbits into the Australian Desert and grasslands. A few soon multiplied into half a billion strong. They ate everything growing in the pasturelands and then, when there was no more grass to supply moisture, the rabbits mobbed the water holes. Scientists got rid of vast numbers by introducing rabbit diseases in 1950, but some hardy animals survived and are still multiplying.

PACKING IT AWAY

You've probably heard of packrats, those rodents that collect almost anything—seeds, bones, rocks—and store it away in their dens. Some of those dens need housecleaning so badly that, over the years, the stuff becomes glued together. What makes it worse is that many generations of packrats often live in the same spot, and the piles can grow for thousands of years. But now the packrats' messy lifestyle may actually be helping scientists. When the burrows are studied, scientists can find out what an area looked like thousands of years ago.

(tamed) and are used by humans for transportation and other needs. Their ability to travel long distances and carry up to 600 pounds (270 kilograms) makes them useful.

HYENA Hyenas are members of the dog family and have one of the strongest jaws of any animal. They range over the deserts of Africa and Arabia, hunting in packs for antelope and other game, or stealing the meal of some other carnivore. They also eat carrion and, occasionally, plant foods. Although they are not particularly fast, they do not give up easily, and may simply wear out their prey, which collapses beneath the snarling pack. Hyenas live as part of a clan in dens they dig themselves.

ENDANGERED SPECIES

As in other biomes, many desert animal species are threatened.

A desert antelope with beautiful horns, the addax has been greatly reduced in numbers by Saharan nomads who believe its stomach contents have healing powers. Its skin, too, is valuable, for it is believed to have the power to ward off attacks from snakes and scorpions.

After World War II (1939–45), when dichlorodiphenyl-trichloroethane (DDT) and other dangerous pesticides were in common use, aplomado falcons shrank in number. These small falcons lived in desert areas of Mexico, Texas, and Arizona, but are no longer found north of Mexico. The poisons were used to kill grasshoppers that were devouring grasses and other vegetation. Grasshoppers are part of the falcons' diet, and as the birds ate the contaminated insects the poison also killed the falcons. It is hoped that the number of falcons will increase now that DDT is banned.

The condor is one of the world's rarest birds. When early settlers slaughtered deer herds for food, they reduced the condor's food supply. Later, cattle ranchers set out poison baits to kill wolves and coyotes, and the poison also killed the condors who ate the dead animals. California condors are now protected, but only a few dozen survive. Some birds have been bred in captivity, however, so there is a chance that the species can be increased.

The beautiful spiral horns of the Arabian oryx are prized by hunters looking for trophies. By 1960 the Arabian oryx were reduced to only about a dozen

---

HOW TEMPERATURES GROW COOLER UNDERGROUND

Burrowing animals escape the heat by going underground, where it can be up to twice as cool as above ground.

Surface land temperature	165°F (74°C)
1 foot (31 centimeters) down	105°F (41°C)
2 feet (61 centimeters) down	95°F (36°C)
3 feet (91 centimeters) down	85°F (29°C)
4 feet (122 centimeters) down	83°F (28°C)

---

animals. In 1961 they became protected and several hundred were bred in zoos. After 1982 many of the zoo-bred animals were reintroduced into the wild.

HUMAN LIFE

Human beings are able to adapt to many unfriendly environments. It is no surprise then, that they have learned to live in the desert and make it their home. Desert dwellers make up less than 1 percent of the world's population, but they include a large variety of native peoples of all races.

IMPACT OF THE DESERT ON HUMAN LIFE

Humans are able to maintain a safe body temperature in the desert by sweating. Under extreme heat, a human being may lose as much as 5 pints (3 liters) of moisture in an hour and up to 21 pints (12 liters) in a day. This water must be replaced, however, or the person will die from dehydration. (Dehydration occurs when tissues dry out, depleting the body of fluids which help keep it cool.) Unlike the kidneys of many desert animals, human kidneys cannot concentrate urine to conserve water. The loss of salt through sweating is also a problem, for humans need a certain amount of salt to maintain energy production. If too much salt is lost, painful muscle cramps can occur.

Because the human animal can change very little physically to adapt to desert conditions, humans must change their behavior. This has been done in many ways, from the choice of lifestyle to the development of technologies such as irrigation (watering of crops). The lifestyles discussed in this chapter are the more traditional ones; most desert peoples today have altered these lifestyles in keeping with the modern world.

Food and water Until the mid-1900s, many native desert dwellers were nomads—either hunter-gatherers, like the Bushmen of the Kalahari and Namib, or herders, like the Bedouin of the Middle East. They moved regularly, usually along an established route, in order to seek food and shelter for themselves or their herds of animals. Most nomads return year after year to the same areas within a given territory. They know what to expect—when the rains usually come, where to find food or pastures, and where a water supply is located. By the late twentieth century, however, less than 3 percent of desert peoples lived nomadic lives, having been driven from their lands by ranchers or mining companies looking for mineral, gas, and oil resources. Since 1950, many nomadic peoples have moved to cities where, too often, they live in poverty.

The diet of hunter-gatherers consists primarily of plant foods and game animals, although meat is usually scarce. Some, like the Aborigines of Australia, eat the grubs (larvae) of certain insects, which provide a source of protein.

Herding tribes depend primarily upon their animals for food, although they may raise grains or trade for them. In the Sahara, dates from the date palm are an important food.

Many Native Americans found important uses for desert plants. The leatherplant, also called sangre de drago (blood of the dragon), contains a red juice used as a medicine for eye and gum diseases. Wine and jelly were made from the fruit of the saguaro cactus, the fruit of the prickly pear was made into jam, and ocotillo branches were useful as building materials.

Shelter The homes of hunter-gatherers tend to be camps, not houses. The Bushmen of the Kalahari, for example, make huts from tree branches and dry grass.

Traditionally, the nomadic tribes of the Sahara, whose shelters had to be portable, used tents made from the hides or hair of herd animals such as goats. The tents of the Bedouin are an example. Village houses in the desert are often made of mud bricks dried in the Sun. Because there is so little rain, the bricks do not need to be waterproof.

In the desert country of Mongolia desert dwellers commonly live in yurts—tentlike structures made from felt created from sheep's wool. The felt is stretched over a wood frame, and a hole at the top of the yurt allows smoke from cooking fires to escape.

Traditionally, some Native Americans of the southwestern desert—the Hopi and Zuni tribes—built homes, called pueblos, from mud, wood, and stone. The thick walls and small windows kept the interior cool. One group, called the Anasazi, who lived around A.D. 1100, built their homes in the sides of cliffs. The homes were reached by ladders, which could be pulled up for defense. The Navajos made hogans, houses of logs and mud. Today, most Native Americans live in the same kinds of homes as other Americans.

Clothing One advantage humans have over other animals is clothing, which is a substitute for fur. Unlike fur, clothing can be put on or taken off at will. Most traditional desert peoples wear layers of loose-fitting garments that actually protect the body from the heat. A naked person absorbs twice as much heat as a person in lightweight clothes. Also, loose clothing absorbs sweat and, as the air moves through, it produces a cooling effect. As a result,

THE WEB OF LIFE: NATURAL BALANCE

Balance is important to every biome. Changes made by humans with even the best of intentions can create serious problems. During the nineteenth century, early settlers brought the American prickly pear cactus to Australia because they liked the plant's appearance. In America the prickly pear is a useful desert shrub, often serving as fencing for livestock. In Australia, where it had no natural enemies, it was a disaster. By 1925, prickly pear cacti covered more than 100,000 square miles (260,000 square kilometers) of the Queensland and Victoria provinces.

Nobody wanted the job of cutting down all those plants. Instead, a natural predator, a little moth that in caterpillar form loves to munch on prickly pears, was brought to Australia. Five years later, no more prickly pears.

the person sweats less, which conserves water. Today, however, many desert tribes, especially in the Middle East, have at least partially adopted western clothing styles.

The only desert peoples to go naked were the Bushmen of the Kalahari and the Aborigines of Australia. Occasionally, when nights were chilly, they wore "blankets" made from bark, but more often their warmth came from a campfire. Those Bushmen and Aborigines who still live a traditional lifestyle continue to go without clothes.

Some kind of headgear is usually worn by desert dwellers to shield the face from the Sun and blowing sand. The Fulani, who live on the edge of the Sahara in West Africa, wear decorated hats made from plant fibers and leather.

Economy For traditional hunter-gatherers, possessions are almost meaningless. If they favor a particular stone for sitting on, they might carry it along. However, a stone, as well as many other things, gets very heavy after a few miles. Their economy tends to be simple and little trading is done.

> ## CAN DESERTIFICATION BE REVERSED?
> The answer to that question in some cases might be yes, because nature fights back. Grasslands in Northern Uganda, Africa, had been lost when herds of cattle were allowed to overgraze the area. Then the tsetse flies moved in and attacked the cattle. When the cattle herders moved to avoid the flies, the plants returned.

A century ago, desert herders were self-sufficient. Their wealth moved with them in the form of herd animals, jewelry, tents, and other possessions. Commerce usually involved selling goats, camels, or cattle. In the Middle East, the discovery of oil made important changes in the economy. In some cases, sweeping modernization made irrigation and food production more stable. Nomads settled in one spot and became farmers. In other cases, the wealth fell into the hands of a few, while the large majority lived in poverty.

IMPACT OF HUMAN LIFE ON THE DESERT

The fact that the desert is so unfriendly to human life has helped preserve it from being overrun by those who could destroy its ecological balance.

Use of plants and animals As long as traditional lifestyles remained in effect, the human impact on the desert was not severe. Desert dwellers understood the need to maintain balance between themselves and their environment. While animals and plants were used for food, they were not exploited (overused), and their numbers could recover. Since the introduction of firearms and the rapid growth of human populations, however, many plants and animals have become threatened.

Desertification, which mean the loss of plant life, continues in spite of efforts by conservationists (people who wish to preserve the environment) to stop it. Several popular desert plants such as cacti, and animals such as lizards, are sold at high prices to collectors. Many of these species are threatened as a result.

Natural resources Overpopulation has diminished many natural resources. The digging of wells has caused the water table (the level of ground water) to drop in many desert regions. Supplies of oil and minerals are being removed from beneath the desert surface and cannot be replaced.

For thousands of years, crops have been grown in desert soil with the aid of irrigation (mechanical watering systems). Furrows were dug between rows of plants, and water pumped from wells and allowed to run along the furrows. In modern times, dams and machinery are used to control the rivers or pump groundwater for irrigation, allowing many former nomads to become settled farmers. To irrigate large cotton farms in the Kara Kum Desert of central Asia, for example, water is brought from the Amu Darya River by means of a canal 500 miles (800 kilometers) long.

However, irrigation must be controlled. If too much groundwater is pumped, it may be used up faster than it can be replaced. When the water table drops in regions near the ocean, the land may slump. Salt water may enter aquifers (underground layers of earth that collects water), destroying the fresh water. Another problem irrigation can cause is an accumulation of salt in the soil. All soils contain some salts and, if irrigation water is used without proper drainage, the salt builds up within the surface layers and plants will no longer grow there.

> ## XERISCAPING
>
> Green lawns and many popular trees and shrubs require constant watering in the summer. This puts a strain on water supplies. Now, some desert communities are asking residents to forget the lawns and xeriscape instead. Xeriscaping is a kind of landscaping using desert plants that need less water.

Quality of the environment People have impacted the desert environment in several ways. Drilling for oil and mining for other resources requires roads. The people who operate the drills need houses. Most of these changes have occurred along the Mediterranean in mineral-rich countries. However, the deep centers of deserts have usually not been disturbed. There, roads remain tracks and have escaped being blacktopped.

The world's climate may be changing because of human activity. If so, the climate of the desert will change as well, and no one knows for sure what that will mean.

DESERT PEOPLES

The Bushmen and the Tuareg are two groups of desert peoples commonly found living in the desert.

Bushmen The Bushmen of the Kalahari and Namib Deserts of Africa live in clans consisting of several families. A clan's territory is about 400 square miles (1,036 square kilometers). Clans move according to the rains or the seasons, returning to familiar campgrounds year after year, and their territory includes good waterholes. Bushmen live off the land, eating berries, roots, and wild game. Plants, which make up the greatest share of their diet, are gathered by the women. Bushmen are expert trackers and use these skills to hunt game for food with bows and poisoned arrows. The meat is cut into strips and dried so that it will keep. Bushmen are not tall, ranging between 55 and 63 inches (140 and 160 centimeters) in height. This may be partially due to a diet deficient in some nutrients.

Overnight shelters built from grass and branches provide protection from the wind. In the winter, clans may break up into smaller groups and build stronger huts that will keep out the rains. Today, the Bushmen number approximately 20,000. Probably fewer than half of those still live as hunter-gatherers.

The Tuareg The northern Tuareg of the Sahara depend upon the camel for their livelihood, grazing their animals on what little pasture exists on the desert fringes. Camels may be killed for meat or their milk used to make butter and cheese. They are also the primary means of transporta-

A Bushman of the Kalahari Desert hunting food with a bow and arrow. (Reproduced by permission of Corbis. Photograph by Anthony Bannister.)

tion. Since the Tuareg are traders, camels are essential to carry goods such as cloth and dates.

The southern Tuareg tribes are more settled. In recent years, as a result of technology that allows the digging of deep wells, the Tuareg have established cattle ranches on the edges of the Sahara. However, the land is often overgrazed and more territory is lost to the desert every year.

Tuareg men wear a characteristic blue veil wound into a turban (head covering) on their heads. Both men and women wear loose robes for protection against the Sun. An indigo (blue) dye is used to color some clothing. Because it often rubs off on the skin, the Tuareg have been known as the "blue men."

Homes are usually low tents made of animal hides dyed red. The tent roof is supported by poles and the sides are tied to the ground with ropes to keep out the wind and sand. During the hottest parts of the year, the Tuareg build a *zeriba,* a large, tall hut made from grasses attached to a wooden frame.

The Tuareg have always been known for their ability to fight. Before firearms were introduced during the eighteenth century, they made impressive weapons such as daggers and swords. They now, however, prefer high-powered rifles. Historically, the Tuareg have had a tendency to raid other tribes, and wars between one tribe and another may last for years.

It is estimated that the Tuareg number as many as 500,000 people. This figure is only approximate, however, since the Tuareg are nomads and are rarely in one place long enough for a reliable count to be made.

> ## EXPERT TRACKERS
>
> Because they are hunters, the Bushmen and the Aborigines are also excellent trackers. Bushmen can follow animal signs over the hardest ground and can distinguish the signs of one individual animal from those of another. During World War II (1939–45), when pilots whose planes had crashed were lost in the Australian deserts, Aborigines could find them by following footprints no one else could see.

THE FOOD WEB

The transfer of energy from organism to organism forms a series called a food chain. All the possible feeding relationships that exist in a biome make up its food web. In the desert, as elsewhere, the food web consists of producers, consumers, and decomposers. The following shows how these three types of organisms transfer energy to create the food web within the desert.

Green plants are the primary producers. They produce organic materials from inorganic chemicals and outside sources of energy, primarily the Sun. Desert annuals and the hardy perennials, such as cacti and palms, turn energy into plant matter.

Animals are consumers. Plant-eating animals, such as locusts, gazelles, and rabbits, are the primary consumers in the desert food web. Secondary consumers eat plant-eaters. They would include the waterhole tadpoles that develop a taste for smaller family members who still eat only plant life. Tertiary consumers are meat-eating predators, like mongooses, owls, and coyotes, which will eat any prey small enough for them to kill. Humans, like the Bushmen, fall into this category. Humans are omnivores, which means they eat both plants and animals.

Decomposers, which feed on dead organic matter, include plants like fungi and animals like the vulture. In moister environments, bacteria also help in decomposition, but they are less effective in the desert's dry climate.

SPOTLIGHT ON DESERTS

THE GOBI DESERT

In the eastern part of central Asia, extending into Mongolia and western China, is the great Gobi Desert—part of a chain of deserts, including the

Nomadic Tuareg in traditional clothing riding camels in the Sahara Desert. (Reproduced by permission of Corbis. Photograph by Tiziana and Gianni.)

Kara-Kum, the Kyzyl-Kum, the Takla Makan, the Alashan, and the Ordos. The name "Gobi" is a Mongolian word meaning "waterless place."

Surrounded by mountains—the Pamirs in the west, the Great Kingan in the east, the Altai, Khangai, and Yablonoi in the north, and the Nan Shan in the south—the Gobi is a high, barren, gravelly plain where few grasses grow. It is so flat that a person can see for miles in any direction

Except for the polar deserts, the Gobi is the coldest because of its altitude, which is about 3,000 feet (900 meters) in the east and about 5,000 feet (1,500 meters) in the south and west. Also, arctic winds blow down from the north so that, in the winter, some areas become snow covered. Any rainfall occurs in the spring and fall. Temperatures average between -40°F (-40°C) in January to 115°F (46°C) in July.

Several rivers flow into the Gobi from the surrounding mountains, but the desert has no oases (waterholes). Although dry river beds and signs of old lakes indicate that water once existed here, the only surface water that remains is alkaline (containing too many undesirable minerals). Fresh water is obtained primarily from wells.

> ## THE GOBI DESERT
> **Location:** Mongolia and western China
> **Area:** 500,000 square miles (1,300,000 square kilometers)
> **Classification:** Cold; arid and semiarid

The few trees found here are willow, elm, poplar, and birch. In general, the largest plants are the tamarisk bush and the saxoul. Grass, thorn, and scrub brush also survive, and bush peas, salt bush, and camel sage are abundant. During the brief spring, annuals (plants that live only one season) flourish.

Reptiles, which do not favor the cold winters, are few; however, several species of snakes live in the Gobi. Many birds, including the sand grouse, the bustard, the eagle, the hawk, and the vulture, thrive here, although many of these are migrators (animals that have no permanent home). Small mammals include jerboas, hamsters, rats, and hedgehogs. Larger mammals include gazelles, sheep, and rare Mongolian wild horses.

The Mongolian people living in the Gobi include settled farmers, who cultivate grains on its fringes, and herders who prefer life on the windy plain. Herders tend to be nomads (wandering tribes), who raise horses, asses, camels, sheep, goats, cattle, and, in upland areas, yaks. The two groups often trade, the farmers providing grain, the herders meat. Since the 1940s, irrigation has been used to make cotton and wheat crops possible, and many nomadic peoples have settled on farms.

Famous Venetian explorer Marco Polo (1254–1324) reached the Gobi during his travels in the thirteenth century. Modern explorers, including archeologists and anthropologists, still visit in search of dinosaur bones and eggs commonly found there. Such fossils are evidence that the Gobi was once a friendlier environment supporting a diverse animal life.

THE THAR DESERT

The Thar (TAHR) Desert (also called the Great Indian Desert) begins near the Arabian Sea and extends almost to the feet of the Himalaya Mountains. Although it contains a variety of landforms and differing climates, it is generally considered an arid, lowland desert characterized by extreme temperature variations. January temperatures average 61°F (16°C), while June temperatures average 99°F (37°C).

The plain of the Thar slopes gently from more than 1,000 feet (305 meters) in the southeast to less than 300 feet (91 meters) in the northwest near the Indus River, broken only by areas of sand dunes, primarily seif (SAFE) dunes, which run parallel to the wind. Some more or less permanent dunes may soar to heights of 200 feet (61 meters). Several salt lakes lie on the desert's margins. The salt covering the lake beds is mined and sold.

Rainfall is unreliable and in the west drops to about 4 inches (100 millimeters) a year. In the summer, the Thar benefits from seasonal monsoons (rainy periods) that visit India, arriving in July and August. During the monsoon season, the largest salt lake may cover 90 square miles (35 square kilometers) and be 4 feet (1.22 meters) deep, only to disappear during the dry period.

Wheat and cotton have been grown on the irrigated Indus plain in the northeast since the 1930s, but most areas support only grass, scrub, jujube,

Aerial view showing the landscape and sand dunes of the Thar Desert. (Reproduced by permission of Corbis. Photograph by Brian Vikander.)

and acacia, which are eaten by domestic camels, cattle, goats, and sheep. Birds, such as the great bustard and quail, are found here. Wild mammals live in or on the fringes of the Thar, including hyenas, jackals, foxes, wild asses, and rabbits.

Nomads raise sheep, cattle, and camels. Villages are located in areas where grass will grow after a rain and support the herds of livestock. Small industries are based on wool, camel hair, and leather.

About 5,000 years ago, the Indus Valley was home to the great Indus civilization. Why this civilization declined and disappeared is still a mystery, although it may have partly resulted from climatic changes that caused the Thar to spread.

THE THAR DESERT

Location: India and Pakistan

Area: 77,000 square miles (200,000 square kilometers)

Classification: Cold; arid

THE AUSTRALIAN DESERT

Most of Australia—about three-quarters of the continent—is desert or semiarid land, a mixture of stone, rock, sand, and clay. January temperatures average 79°F (26°C), and July temperatures average 53°F (12°C). The central region is a basin, the western area a high plateau. Although regions within the Australian Desert have different names—the Simpson, the Victoria, the Gibson, and the Great Sandy—they blend into one another and are part of the same whole.

THE AUSTRALIAN DESERT

Location: Australia

Area: 600,000 square miles (1,500,000 square kilometers)

Classifications: Hot; arid and semiarid

The basin region is a flat land, where a visitor can gaze for miles without the view being interrupted by hills. This makes it a difficult place in which to navigate, for there are no landmarks. However, the terrain varies, changing from semiarid grassland to rocky stretches, to areas with sandy dunes. Strange, massive, solitary rock formations, which have spiritual meaning to the Aborigines—such as Ayers Rock or the Olgas—stand like giant sentinels.

The lowest point of the Australian Desert is near Lake Eyre, which lies 50 feet (15 meters) below sea level. The lake, situated close to sand dunes, is about 50 miles (80 kilometers) long and so salt-encrusted that during dry periods it appears white. It fills with water only once or twice during a ten-year period, during which time it supports abundant wildlife, including birds, frogs, and toads.

Mulga, a type of thorny acacia bush, and mallee, a species of eucalyptus, grow in the most arid regions of the Australian Desert. At the desert's edges, spinifex grass grows.

Although Australia is home to a number of animals found nowhere else in the world, their ways of adapting to the desert are the same as animals elsewhere. Insects, such as locusts and termites; snakes, such as the bandy bandy and the brown snake; and lizards, such as the blue-tongued skink are commonly found in the Australian Desert. Birds include the emu, a large, flightless bird that resembles an ostrich, and parrots, quail, cockatoos, and kookaburras.

Kangaroos are Australia's largest mammals. Although they prefer to live in grasslands, they can travel long distances in the desert without water. Kangaroos are marsupials and carry their young in a pouch. Other marsupials unique to Australia are the wallaby, the wombat, and the Tasmanian wolf.

Until the middle of the twentieth century, most Aborigines of Australia still lived as hunter-gatherers. Traveling in small groups, or clans, they moved over an established territory. They gathered plant roots, bulbs, termites, and grubs; and hunted kangaroos and wallabies with spears and boomerangs. As young Aborigine men attained a certain age, they went on a "walkabout." This meant they had to leave their clan to wander in the desert, perhaps for years, learning about life and survival.

THE ARABIAN DESERT

The Arabian Desert covers most of the Arabian Peninsula, from the coast of the Red Sea to the Persian Gulf. Except for fertile spots in the south-

The oblonged-shaped Ayers Rock found in the Australian Desert. (Reproduced by permission of Popperfoto/Archive Photos, Inc.)

east and southwest, the peninsula is all desert. During prehistoric times, volcanic cones and craters were formed along its western edge. On the eastern side, sedimentary rocks and prehistoric sea life formed the world's richest oilfields. Inside some of these same rocks are vast supplies of underground water, captured during ages past when the area included wetlands.

In the south are deep ravines. Otherwise, the desert plateau consists of bare rock, gravel, or sand having a characteristically golden color. Wind and occasional flooding has carved the rocks into fantastic shapes. The Empty Quarter—about a third of the southern part of the peninsula—is a vast sea of sand dunes, some 1,110 feet (330 meters) high and 125 miles (200 kilometers) long. Treacherous quicksand can also be found here, its particles so smooth that they act like ballbearings, drawing unwary creatures below the surface to their deaths. Another area of dunes, the Great Nafud, is found farther north and contains so few waterholes that even camels find it difficult to cross. Along the coasts, changing sea levels have resulted in large salt flats, some as much as 20 miles (30 kilometers) wide.

A common wombat, a squirrel or rabbit-like creature, found in the Australia. (Reproduced by permission of WWF/Klein-Hubert.)

Although a monsoon season (rainy period) occurs along the southeast portion, rainfall elsewhere is scanty—only 3 to 4 inches (77 to 102 millimeters) a year. Flash floods are common during the infrequent rains, and hailstorms are not unknown. Droughts may last several years. January temperatures average 65 °F (18 °C).

In the southern portion of the Arabian Desert annuals spread a colorful blanket over the soil right after a rain; otherwise the hardy perennials are the most common plant life, including mimosas, acacias, and aloe. Oleanders and some species of roses also thrive. Few trees can survive; however, the tamarisk tree helps control drifting sand, and junipers grow in the southwest. Date palms grow almost everywhere except at high elevations, and coconut palms can be found on the southern coast. Plants which can be cultivated with the aid of irrigation include alfalfa, wheat, barley, rice, cotton, and many fruits, including mangoes, melons, pomegranates, bananas, and grapes.

Swarms of locusts move over the land periodically, causing much destruction. Other common invertebrates include ticks, beetles, scorpions, and ants. Horned vipers and a special species of cobra make their home in the Arabian Desert, as do monitors and skinks. Ostriches are now extinct there, but eagles, vultures, and owls are common. Seabirds, such as pelicans, can be seen on the coasts. Wild mammals include the gazelle, oryx, ibex, hyena, wolf, jackal, fox, rabbit, and jerboa. The lion once lived there but has long been extinct.

> **THE ARABIAN DESERT**
>
> **Location:** The Arabian Peninsula
>
> **Area:** 900,000 square miles (2,300,000 square kilometers)
>
> **Classification:** Hot; extremely arid and arid

For centuries, the Arabian Desert has been home to nomadic tribes of Bedouin. The camel makes life in the desert possible for them. A camel's owner can live for months in the desert on the camel's milk. The camel is also used for meat, clothing, and muscle power, and its dung (solid waste) is burned for fuel. Domestic sheep and goats are raised, as well as donkeys and horses.

THE SAHARA DESERT

The Sahara ranges across the upper third of Africa, from the Atlantic Ocean to the Red Sea, and is about 1,250 miles (2,000 kilometers) wide from north to south. It is the world's largest desert, covering an area almost as large as the United States. Its landforms, which tend to have a golden color, range from rocky mountains and highlands—some as high as 11,000 feet (3,300 meters)—to stretches of gravel and vast sand dunes. In places, dunes may reach 1,000 feet (305 meters) in height. Erosion has shaped the sandstone rocks in some areas into weird and wonderful shapes and deep, narrow canyons.

> **THE SAHARA DESERT**
>
> **Location:** North Africa
>
> **Area:** 3,200,000 square miles (8,600,000 square kilometers)
>
> **Classification:** Hot; extremely arid and arid

Millions of years ago, volcanic activity occurred here. The region was also the site of shallow seas and lakes, which contributed to the vast reserves of oil deposits now found. Around 150 B.C., when the Romans controlled North Africa, the northern Sahara was a rich agricultural area producing about 500,000 tons (453,500 metric tons) of wheat each year. Over the centuries, however, sand has claimed the once fertile landscape.

The Sahara is the hottest desert, with a mean annual temperature of 85°F (29°C). Nights are cool, often as much as 60°F (33°C) cooler than the days. Except along the southern fringe, rain is not dependable, and may be absent for as much as 10 years in succession. Then it tends to fall in sudden storms.

In some areas, such as the Tanezrouft region, nothing appears to grow. Elsewhere, annuals bloom after the unpredictable rains and provide food for camels and wild animals. At one time, the only perennial vegetation at oases were tamarisks and oleander bushes. Then the date palm was introduced and now citrus fruits, peaches, apricots, wheat, barley, and millet are all cultivated.

Animals such as gazelles, oryx, addax, foxes, badgers, and jackals live in the wild. Domestic animals include camels, sheep, and goats.

Three main groups of people now live in the Sahara: the Tuareg, the Tebu, and the Moors. Two-thirds of the population live at oases, where they depend upon irrigation and the deep wells that tap underground water.

THE PATAGONIAN DESERT

Along the entire length of Argentina, between the Andes Mountains in the west and the Atlantic Ocean in the east, lies the Patagonian Desert. It owes its existence to a cold ocean current and the Andes Mountains, both of which cause dry air to form. The terrain is a series of plateaus, some as high as 5,000 feet (900 meters) near the Andes, which slope toward the sea. Deep, wide valleys made by ancient rivers have created clifflike walls, but only a few streams remain. These are usually formed by melting snow from the Andes.

THE PATAGONIAN DESERT

Location: Argentina

Area: 260,000 square miles (673,000 square kilometers)

Classification: Cold; arid

In some areas, the soil is alkaline or covered by salt deposits and does not support plant life. Elsewhere it is primarily gravel. Mineral resources, including coal, oil, and iron ore, have been found in several places. However, quantities are too small to be important.

Rainfall, which occurs in the summer, is usually less than 8 inches (20 centimeters) annually. Although arid, the Patagonian Desert does not experience extremes of temperature, primarily because it is so close to the ocean. Even during the warmest months, temperatures rarely exceed 54°F (12°C), and during the colder months, subzero temperatures and snow are common. Mean annual temperatures range from 54° to 68°F (12° to 20°C) in the north and 43° to 55°F (6° to 13°C) in the south. It is also windy and winds of 70 miles (112 kilometers) per hour are not uncommon.

Vegetation consists of tough perennial bushes and clumps of grass on which sheep graze. There are no trees.

Many reptiles are found here, but small mammals are the most numerous animals. Rabbits and hares are widespread. The rhea, an American ostrich, lives throughout the region. Other common animals are the mara and the guanaco, a llama-like animal prized for its long, fine hair. Because of the fine quality of its coat, the guanaco was hunted almost to extinction.

The most unusual animal of the Patagonian Desert is the armadillo. Its skin consists of plates of tough armor, including one that covers its face. It can travel very fast on its short legs and, when threatened, digs a hole to escape.

The name "Patagonia" means "big feet," and refers to the Tehuelche Indians who, when first seen by Portugese explorer Ferdinand Magellan (c. 1480–1521) in 1520, were wearing oversized boots. The Tehuelche were nomadic hunters whose size and vigor caused Europeans to consider them giants. After the Europeans introduced the Tehuelche to horses, their ability to seek new territory and intermarry with other tribes increased. By 1960, the race was virtually extinct. Modern settlements in and around the Patagonian Desert were established during the twentieth century, after oil was discovered on the coast.

THE ATACAMA DESERT

The Atacama Desert is a long, narrow, coastal desert that is seldom more than 125 miles (200 kilometers) wide. The Pacific Ocean lies to the west, where sheer cliffs about 1,475 feet (450 meters) high rise from the sea. Beyond these cliffs lies a barren valley which runs along the foothills of the Andes Mountains.

Cold ocean currents are responsible for the dry conditions which make the Atacama the world's driest area. Droughts may persist for many years, and there is no dependable rainy season. Although it is classified as a hot desert, mean temperatures seldom exceed 68°F (20°C).

> **THE ATACAMA DESERT**
> **Location:** Chile and parts of Peru
> **Area:** 54,000 square miles (140,000 square kilometers)
> **Classification:** Hot; extremely arid and arid

Much of the Atacama consists of salt flats over gravelly soil, although sand dunes have formed in a few areas. In the south, a raised plateau nearly 3,300 feet (1,006 meters) high and broken with volcanic cones takes on an otherworldly appearance. In the southeast, a plateau bordering the Andes reaches a height of 13,100 feet (4,000 meters).

Although the area is rich in boron, sodium nitrate, and other minerals, they contaminate any underground water, making it unusable both by plants and animals. Plant life consists primarily of coarse grasses, mesquite, and a few cacti.

Animal life is also rare. Perhaps the most commonly found animals are lizards, although they are not numerous. The rather timid giant iguanas, some growing over 6 feet (2 meters) long and resembling dragons prowling are scavengers. Huge condors, which feed on carrion, can be seen soaring overhead. Wherever cacti are found, cactus wrens feed and breed, and oven birds, which build their own little adobe (mud) huts instead of nests, live in the less arid regions. The only mammals living in the Atacama are small rodents, such as the chinchilla, although guanaco and vicuña inhabit higher elevations.

Indians may have lived in or on the fringes of the Atacama at one time; however, they may have been killed by European settlers during the eighteenth century. Those people who live in the area today are descendants of the European immigrants.

The Atacama was once important to the fertilizer industry for which sodium nitrate was mined. Another fertilizer, bird droppings called guano

A highway cuts through the arid hills of the Atacama Desert in northern Chile. (Reproduced by permission of Corbis. Photograph by Jeremy Horner.)

(GWAN-oh), was collected on the offshore islands, which are breeding areas for seabirds. In the 1920s, synthetic fertilizers became popular, and the mining settlements in the Atacama were abandoned.

THE NAMIB DESERT

A coastal desert, the Namib is part of the vast tableland (flat highland) of southern Africa. Although it borders the Kalahari Desert to the east, the two are very different. Some scientists do not consider the Kalahari a true desert, while the Namib certainly is one. The terrain is gravelly in the north, a rich area for gemstones, particularly diamonds. In the south sand prevails, creating enormous dunes as much as 1,000 feet (305 meters) high in some areas.

Rain is rare, and years may pass between showers. When rain does come, however, it creates flash floods and then disappears as quickly as it came. Although winds blow in from the Atlantic, they do not bring storms. Instead, dense fog rolls in almost nightly, creating very humid air.

Unusual life forms exist in the Namib, particularly in the dunes in the south, where plant life depend upon the fog and has evolved to draw moisture from it. The strange welwitschia plant is an example.

> **THE NAMIB DESERT**
>
> **Location:** Eastern Namibia
>
> **Area:** 52,000 square miles (135,000 square kilometers)
>
> **Classification:** Hot; arid

A species of zebra that lives in the Namib is able to sniff out small pools of water that may lie in gullies or dry stream beds. Then the zebras dig, sometimes 2 or 3 feet (.6 or .9 meter) into the ground, until the water is uncovered.

Although Bushmen live principally in the Kalahari, they frequent the Namib as well.

THE MOJAVE DESERT

Deserts in North America are all part of the Great Basin, a vast territory stretching from just north of the Canadian border into parts of Mexico. In the north, cold semiarid deserts have formed, ringed by the Sierra Nevada and Rocky Mountains. Further south lie the hot deserts, the Mojave and the nearby Sonora Desert.

The Mojave Desert, named for the Mojave Indians who once lived along the Colorado River, consists of salt flats, barren mountains, deep ravines, high plateaus, and wide, windy plains of sand. Elevations range from 2,000 to 5,000 feet (600 to 1,500 meters). During prehistoric times, the

> **THE MOJAVE DESERT**
>
> **Location:** Southeastern California, Nevada, Arizona, Utah
>
> **Area:** 25,000 square miles (65,000 square kilometers)
>
> **Classification:** Hot; arid

Pacific Ocean covered the area until volcanic action built the mountain ranges, leaving salt pans and mudflats as the only remaining signs. The desert is rich in minerals. Borax, potash, salt, silver, and tungsten are mined here.

Annual rainfall is less than 5 inches (13 centimeters). Frost is common in the winter, and snow occasionally falls. One major river, the Mojave, crosses the desert, running underground for part of its length. July temperatures average 75°F (24°C), and winter temperatures often drop below freezing.

Vegetation characteristic to North American deserts include different species of cacti such as organ pipe, prickly pear, and saguaro. Although nothing grows on the salt flats, plants, such as the creosote bush, can find a foothold in other areas. The rare Joshua trees are protected in the Joshua Tree National Park. On the highest mountains, piñon and juniper grow.

A wide variety of invertebrates, amphibians, reptiles, birds, and smaller mammals make a home in this region. Larger mammals, such as the puma, jaguar, peccary, prong-horned antelope, and bighorn sheep, are also found.

DEATH VALLEY

Death Valley is a desert basin of which approximately 550 square miles (1430 square kilometers) lie below sea level. Near its center is the lowest point in North America, -282 feet (-86 meters). Prehistoric salt flats exist in the lowest areas, where nothing grows.

It is also the hottest place on the continent, summer temperatures often exceed 120°F (53°C). In 1913, a temperature of 134.6°F (61.6°C) was recorded. Rainfall is seldom more than 2 inches (50 millimeters) per year. However, many species of plants are found here. Annuals, such as poppies, appear in late winter and early spring; perennials, such as cacti and mesquite, survive all year. Lizards, foxes, rats, mice, squirrels, coyotes, bighorn sheep, wild burros, and rabbits live here, as well as many birds, such as ravens.

> **DEATH VALLEY**
> **Location:** California
> **Area:** 5,312 square miles (13,812 square kilometers)
> **Classification:** Hot; arid

Death Valley got its name from the numbers of gold-seekers who died there during the gold rush in the mid-1800s. Although gold, silver, lead, and copper have been mined in the area, Death Valley became known for borax (compound used as an antiseptic and cleanser), discovered in 1873 and brought out by mule teams. Mining ghost towns now draw tourists, and Death Valley became a national park in 1933.

Perhaps Death Valley's most famous inhabitant was "Death Valley Scotty." His real name was Walter Edward Scott (1872–1954), a cowboy in Buffalo Bill's Wild West Show. In 1905, Scott built "Scotty's Castle," a $2,000,000 home in Grapevine Canyon in Death Valley, where he lived for more than 30 years.

THE ANTARCTIC DESERT

Polar deserts are like other deserts only in the dryness of the air. Here, all moisture is frozen. Temperatures are very different. In the polar desert, the temperature rarely goes above freezing. Mean temperature during warmest month is from -30° to -4°F (-34° to -20°C), and during coldest month is -94° to -40°F (-70° to -40°C).

Landforms may also be different. Scientists still know very little about what the land looks like in Antarctica because it is buried beneath such deep ice—over 14,000 feet (4,270 meters) deep in places. It is believed, however, that the area is a series of islands connected by great ice sheets. Western Antarctica is mountainous, and eastern Antarctica is a plateau (high, flat land).

> ## THE ANTARCTIC DESERT
> **Location:** Antarctica
> **Area:** Approximately 360,000 square miles (936,000 square kilometers)
> **Classification:** Polar

Tiny communities of microbes (microorganisms) have been found in the ice desert of Antarctica. During the summer, the Sun warms little pockets of dirt and grit and turns them to slush. In the dim light, bacteria suspended in the slush can photosynthesize.

Plant growth is limited to the bordering tundra regions where lichens and mosses are the largest plant forms. Algae, yeasts, lower fungi, and bacteria are also found in some areas. Grasses and a few other seed plants may grow on the fringes. The driest areas support no known plant life.

The only insects that live in Antarctica appear to be parasites that live on birds and mammals. These include lice, mites, and ticks.

Seabirds, such as petrels and terns, frequent the coast where they feed on fish. However, the only true Antarctic bird is the emperor penguin. These penguins live in large flocks and nest in the autumn, living on fish they catch in the coastal waters.

Marine mammals, such as whales and seals, live in the coastal waters, but they do not enter desert areas. (Polar bears live only at the North Pole; they do not live in Antarctica.)

Temporary year-round human settlements were established in Antarctica around 1900, primarily for exploration purposes. Any industry is centered on the surrounding sea, primarily whaling and seal hunting.

FOR MORE INFORMATION

BOOKS

Arritt, Susan. *The Living Earth Book of Deserts*. New York: Reader's Digest, 1993.

The Hospitable Desert. Hauppauge, NY: Barron's Educational Series, Inc., 1999.

Lambert, David. *People of the Desert*. Orlando, FL: Raintree Steck-Vaughn Publishers, 1999.

Plants of the Desert. Broomall, PA: Chelsea House Publishers, 1997.

Savage, Stephen. *Animals of the Desert.* Orlando, FL: Raintree Steck-Vaughn Publishers, 1996.

ORGANIZATIONS

Chihuahuan Desert Research Institute
  PO Box 905
  Fort Davis, TX 79734
  Phone: 915-364-2499; Fax: 915-837-8192
  Internet: http://www.cdri.org

Desert Protective Council, Inc.
  PO Box 3635
  San Diego, CA 92163
  Phone: 619-298-6526; Fax: 619-670-7127

Environmental Defense Fund
  257 Park Ave. South
  New York, NY 10010
  Phone: 800-684-3322; Fax: 212-505-2375
  Internet: http://www.edf.org

Environmental Network
  4618 Henry Street
  Pittsburgh, PA 15213
  Internet: http://www.envirolink.org

Environmental Protection Agency
  401 M Street, SW
  Washington, DC 20460
  Phone: 202-260-2090
  Internet: http://www.epa.gov

Friends of the Earth
  1025 Vermont Ave. NW, Ste. 300
  Washington, DC 20003
  Phone: 202-783-7400; Fax: 202-783-0444

Greenpeace USA
  1436 U Street NW
  Washington, DC 20009
  Phone: 202-462-1177; Fax: 202-462-4507
  Internet: http://www.greenpeaceusa.org

Sierra Club
  85 2nd Street, 2nd fl.
  San Francisco, CA 94105
  Phone: 415-977-5500; Fax: 415-977-5799
  Internet: http://www.sierraclub.org

World Wildlife Fund
  1250 24th Street NW
  Washington, DC 20037
  Phone: 202-293-4800; Fax: 202-293-9211
  Internet: http://www.wwf.org

WEBSITES

Note: Website addresses are frequently subject to change.

National Center for Atmospheric Research: http://www.dir.ucar.edu

World Meteorological Organizaion: http://www.wmo.ch

BIBLIOGRAPHY

Cloudsley-Thompson, John. *The Desert*. New York: G.P. Putnam's Sons, 1977.

"Desert." *Encyclopaedia Britannica*. Chicago: Encyclopaedia Britannica, Inc., 1993.

"Desert." *Grolier Multimedia Encyclopedia*. New York: Grolier, Inc., 1995.

Dixon, Dougal. *The Changing Earth*. New York: Thomson Learning, 1993.

Flegg, Jim. *Deserts: A Miracle of Life*. London: Blandford Press, 1993.

George, Michael. *Deserts*. Mankato, Minnesota: Creative Education, Inc., 1992.

Leopold, A. Starker. *The Desert*. New York: Time Incorporated, 1962.

Lovelock, James. *The Ages of Gaia: A Biography of Our Living Earth*. New York: Bantam Books, 1990.

Macquitty, Miranda. *Desert*. New York: Alfred A. Knopf, 1994.

Oram, Raymond F. *Biology: Living Systems*. Westerville, OH: Glencoe/McGraw-Hill, 1994.

Sayre, April Pulley. *Desert*. Exploring Earth's Biomes. New York: Twenty-first Century Books, 1994.

Waterlow, Julia. *Deserts*. Habitats. New York: Thomson Learning, 1996.

Watson, Lyall. *Dark Nature*. New York: HarperCollins Publishers, 1995.

Whitfield, Philip, ed. *Atlas of Earth Mysteries*. Chicago: Rand McNally, 1990.

BIBLIOGRAPHY

BOOKS

Aldis, Rodney. *Ecology Watch: Polar Lands*. New York: Dillon Press, 1992.

"America's Wetlands." Environmental Protection Agency, http://www.epa.gov/OWOW/wetlands/vital/what.html

Amos, William H. *The Life of the Seashore*. New York: McGraw-Hill Book Company, 1966.

"Arctic National Wildlife Refuge." U. S. Fish and Wildlife Services, http://www.r7.fws.gov/nwr/arctic/arctic.html

Arritt, Susan. *The Living Earth Book of Deserts*. New York: Reader's Digest, 1993.

Ayensu, Edward S., ed. *Jungles*. New York: Crown Publishers, Inc., 1980.

Baines, John D. *Protecting the Oceans*. Conserving Our World. Milwaukee, WI: Raintree Steck-Vaughn, 1990.

Barber, Nicola. *Rivers, Ponds, and Lakes*. London: Evans Brothers Limited, 1996.

Barnhart, Diana, and Vicki Leon. *Tidepools*. Parsippany, NJ: Silver Burdett Press, 1995.

"Basic Soils." University of Minnesota: http://www.soils.agi.umn/edu/academics/classes/soil3125/index.html

"Beach and Coast." *Grolier Multimedia Encyclopedia*. Danbury, CT: Grolier, Inc., 1995.

Beani, Laura, and Francesco Dessi. *The African Savanna*. Milwaukee, WI: Raintree Steck-Vaughn Publishers, 1989.

Borgioli, A., and G. Cappelli. *The Living Swamp*. London: Orbis, 1979.

Brown, Lauren. *Grasslands*. New York: Alfred A. Knopf, 1985.

Burnie, David. *Tree*. New York: Alfred A. Knopf, 1988.

Burton, John A. *Jungles and Rainforests*. The Changing World. San Diego, CA: Thunder Bay Press, 1996.

Burton, Robert, ed. *Nature's Last Strongholds*. New York: Oxford University Press, 1991.

Campbell, Andrew. *Seashore Life*. London: Newnes Books, 1983.

Capstick, Peter Hathaway. *Death in the Long Grass*. New York: St. Martin's Press, 1977.

Catchpoh, Clive. *The Living World: Mountains*. New York: Dial Books for Young Readers, 1984.

Caras, Roger. *The Forest: A Dramatic Portrait of Life in the American Wild*. New York: Holt, Rinehart, Winston, 1979.

Carroll, Michael W. *Oceans and Rivers*. Colorado Springs, CO: Chariot Victor Publishing, 1999.

BIBLIOGRAPHY

Carwardine, Mark. *Whales, Dolphins, and Porpoises.* See and Explore Library. New York: Dorling Kindersley, 1992.

Cerullo, Mary M. *Coral Reef: A City That Never Sleeps.* New York: Cobblehill Books, 1996.

"Chaparral." Britannica Online.http://www.eb.com:180/cgi-bin/g? DocF=micro/116/70.html

Christiansen, M. Skytte. *Grasses, Sedges, and Rushes.* Poole, Great Britain: Blandford Press, 1979.

Claybourne, Anna. *Mountain Wildlife.* Saffron Hill, London: Usborne Publishing, Ltd., 1994.

Cloudsley-Thompson, John. *The Desert.* New York: G.P. Putnam's Sons, 1977.

Cole, Wendy. "Naked City: How an Alien Ate the Shade." *Time.* (February 15, 1999): 6.

Cricher, John C. *Field Guide to Eastern Forests.* Peterson Field Guides. Boston: Houghton Mifflin, 1988.

Cumming, David. *Coasts.* Habitats. Austin, TX: Raintree Steck-Vaughn, 1997.

Curry-Lindahl, Kai. *Wildlife of the Prairies and Plains.* New York: Harry N. Abrams, Inc., 1981.

Davis, Stephen, ed. *Encyclopedia of Animals.* New York: St. Martin's Press, 1974.

"Desert." *Encyclopaedia Britannica.* Chicago: Encyclopaedia Britannica, Inc., 1993.

"Desert." *Grolier Multimedia Encyclopedia.* New York: Grolier, Inc., 1995.

Dixon, Dougal. *The Changing Earth.* New York: Thomson Learning, 1993.

Dixon, Dougal. *Forests.* New York: Franklin Watts, 1984.

Dudley, William. *Endangered Oceans.* Opposing Viewpoints Series. San Diego, CA: Greenhaven Press, Inc., 1999.

Duffy, Eric. *The Forest World: The Ecology of the Temperate Woodlands.* New York: A&W Publishers, Inc., 1980.

Duffy, Trent. *The Vanishing Wetlands.* New York: Franklin Watts, 1994.

Dugan, Patrick, ed. *Wetlands In Danger: A World Conservation Atlas.* New York: Oxford University Press, 1993.

Engel, Leonard. *The Sea.* Life Nature Library. New York: Time-Life Books, 1969.

Everard, Barbara, and Brian D. Morley. *Wild Flowers of the World.* New York: Avenal Books, 1970.

Everglades: Encyclopedia of the South. New York: Facts on File, 1985.

Feltwell, John. *Seashore.* New York: DK Publishing, Inc., 1997.

Finlayson, Max, and Michael Moser. *Wetland.* New York: Facts on File, 1991.

Fisher, Ron. *Heartland of a Continent: America's Plains and Prairies.* Washington, DC: National Geographic Society, 1991.

Fitzharris, Tim. *Forests.* Canada: Stoddart Publishing Co., 1991.

Flegg, Jim. *Deserts: A Miracle of Life.* London: Blandford Press, 1993.

"Forest." *Colliers Encyclopedia.* CD-ROM, P. F. Collier, 1996.

"Forestry and Wood Production; Forestry: Purposes and Techniques of Forest Management; Fire Prevention and Control." Britannica Online. http://www.eb.com:180/cgi-bin/g? DocF=macro/5002/41/19.html

"Forests." Britannica Online. http://www.eb.com:180/cgi-bin/g?DocF=macro/5002/41/19.html

"Forests and Forestry." *Grolier Multimedia Encyclopedia.* New York: Grolier, Inc., 1995.

Forman, Michael. *Arctic Tundra.* New York: Children's Press, 1997.

Fornasari, Lorenzo, and Renato Massa. *The Arctic*. Milwaukee, WI: Raintree Steck-Vaughn Publishers, 1989.

Fowler, Allan. *Arctic Tundra*. Danbury, CT: Children's Press, 1997.

Fowler, Allan. *Life in a Wetland*. Danbury, CT: Children's Press, 1999.

Franck, Irene, and David Brownstone. *The Green Encyclopedia*. New York: Prentice Hall, 1992.

Ganeri, Anita. *Ecology Watch: Rivers, Ponds and Lakes*. New York: Dillon Press, 1991.

Ganeri, Anita. *Habitats: Forests*. Austin, Texas: Raintree Steck-Vaughn, 1997.

Ganeri, Anita. *Ponds and Pond Life*. New York: Franklin Watts, 1993.

George, Jean Craighead. *Everglades Wildguide*. Washington, DC: National Park Service, U.S. Dept. of the Interior, 1988.

George, Jean Craighead. *One Day in the Alpine Tundra*. New York: Thomas Y. Crowel, 1984.

George, Michael. *Deserts*. Mankato, Minnesota Creative Education, Inc., 1992.

Gilman, Kevin. *Hydrology and Wetland Conservation*. New York: John Wiley & Sons, Inc., 1994.

Gleseck, Ernestine. *Pond Plants*. Portsmouth, NH: Heinemann Library, 1999.

"Glow-in-the-Dark Shark Has Killer Smudge." *Science News* 154 (August 1, 1998): 70.

"Grass." *WorldBook Encyclopedia*. Chicago: World Book Incorporated, Scott Fetzer Co., 1999.

Greenaway, Theresa, Christiane Gunzi, and Barbara Taylor *Forest*. New York: DK Publishing, 1994.

Greenaway, Theresa. *Jungle*. Eyewitness Books. New York: Alfred A. Knopf, 1994.

Grisecke, Ernestine. *Wetland Plants*. Portsmouth, NH: Heinemann Library, 1999.

Gutnick, Martin J., and Natalie Browne-Gutnik. *Great Barrier Reef*. Wonders of the World. Austin, TX: Raintree Steck-Vaughn Publishers, 1995.

Grzimek, Bernhard. *Grizmek's Animal Encyclopedia: Fishes II and Amphibians*. New York: Van Nostrand Reinhold, 1972.

Grzimek, Bernhard. *Grizmek's Animal Encyclopedia: Reptiles*. New York: Van Nostrand Reinhold, 1972.

Hargreaves, Pat, ed. *The Indian Ocean*. Seas and Oceans. Morristown, NJ: Silver Burdett Company, 1981.

Hargreaves, Pat, ed. *The Red Sea and Persian Gulf*. Seas and Oceans. Morristown, NJ: Silver Burdett Company, 1981.

Hecht, Jeff. *Vanishing Life*. New York: Charles Scribner & Sons, 1993.

Hirschi, Ron. *Save Our Forests*. New York: Delacorte Press, 1993.

Hirschi, Ron. *Save Our Prairies and Grasslands*. New York: Delacorte Press, 1994.

Hirschi, Ron. *Save Our Wetlands*. New York: Delacorte Press, 1995.

Hiscock, Bruce. *Tundra: The Arctic Land*. New York: Atheneum, 1986.

Holing, Dwight. *Coral Reefs*. Parsippany, NJ: Silver Burdett Press, 1995.

The Hospitable Desert. Hauppauge, NY: Barron's Educational Series, Inc., 1999.

Inseth, Zachary. *The Tundra*. Chanhassen, MN: The Child's World, Inc., 1998.

International Book of the Forest. New York: Simon and Schuster, 1981.

Kaplan, Elizabeth. *Biomes of the World: Taiga*. New York: Benchmark Books, 1996.

Kaplan, Elizabeth. *Biomes of the World: Temperate Forest*. New York: Benchmark Books, 1996.

BIBLIOGRAPHY

Kaplan, Elizabeth. *Biomes of the World: Tundra*. New York: Marshall Cavendish, 1996.

Kaplan, Eugene H. *Southeastern and Caribbean Seashores*. NY: Houghton Mifflin, Company, 1999.

Khanduri, Kamini. *Polar Wildlife*. London: Usborne Publishing, Ltd., 1992.

Knapp, Brian. *Lake*. Land Shapes. Danbury, CT: Grolier Educational Corporation, 1993.

Knapp, Brian. *River*. Land Shapes. Danbury, CT: Grolier Educational Corporation, 1992.

Knapp, Brian. *What Do We Know About Grasslands?* New York: Peter Bedrick Books, 1992.

"Kola Peninsula." Britannica Online. http://www.eb.com:180/cgi-bin/g?DocF=micro/326/3.html

"Lake." *Encyclopaedia Britannica*. Chicago: Encyclopaedia Britannica, Inc., 1993.

"Lake." *Grolier Multimedia Encyclopedia*. Danbury, CT: Grolier, Inc., 1995.

"Lakes." *Britannica Online*. http://www.eb.com.180/cgi-bin/g?DocF=macro/5003/58.html

Lambert, David. *Our World: Grasslands*. Parsippany, NJ: Silver Burdett Press, 1988.

Lambert, David. *People of the Desert*. Orlando, FL: Raintree Steck-Vaughn Publishers, 1999.

Lambert, David. *Seas and Oceans*. New View. Milwaukee, WI: Raintree Steck-Vaughn Company, 1994.

Leopold, A. Starker. *The Desert*. New York: Time Incorporated, 1962.

Levathes, Louise E. "Mysteries of the Bog." *National Geographic*. Vol 171, No. 3, March 1987, pp. 397–420.

Levete, Sarah. *Rivers and Lakes*. Brookfield, CT: Millbrook Press, Inc., 1999.

Leyder, Francois. "Okefenokee, The Magical Swamp." *National Geographic*. Vol. 145, No. 2, Feb. 1974, pp. 169–75.

Lourie, Peter. *Hudson River: An Adventure from the Mountains to the Sea*. Honesdale, PA: Boyds Mills Press, 1992.

Lovelock, James. *The Ages of Gaia: A Biography of Our Living Earth*. New York: Bantam Books, 1990.

Lye, Keith. *Our World: Mountains*. Morristown, NJ: Silver Burdett, 1987.

Macquitty, Miranda. *Desert*. New York: Alfred A. Knopf, 1994.

Macquitty, Miranda. *Ocean*. Eyewitness Books. New York: Alfred A. Knopf, 1995.

Madson, John. *Tallgrass Prairies*. Helena, MT: Falcon Press Publishing, 1993.

Markle, Sandra. *Pioneering Ocean Depths*. New York: Atheneum, 1995.

Massa, Renato. *Along the Coasts*. Orlando, FL: Raintree Steck-Vaughn Publishers, 1998.

Massa, Renato. *India*. World Nature Encyclopedia. Milwaukee, WI: Raintree Steck-Vaughn Publishers, 1989.

McCormick, Anita Louise. *Vanishing Wetlands*. San Diego: Lucent Books, 1995.

McCormick, Jack. *The Life of the Forest*. New York: McGraw-Hill, 1966.

McLeish, Ewan. *Habitats: Oceans and Seas*. Austin, TX: Raintree Steck-Vaughn Company, 1997.

McLeish, Ewan. *Habitats: Wetlands*. New York: Thomson Learning, 1996.

Moore, David M., ed. *The Marshall Cavendish Illustrated Encyclopedia of Plants and Earth Sciences*. Vol. 7. New York: Marshall Cavendish, 1990.

Morgan, Nina. *The North Sea and the Baltic Sea*. Seas and Oceans. Austin, TX: Raintree Steck-Vaughn Company, 1997.

Morgan, Sally. *Ecology and Environment*. New York: Oxford University Press, 1995.

Nadel, Corinne J., and Rose Blue. *Black Sea*. Wonders of the World. Austin,

TX: Raintree Steck-Vaughn Company, 1995.

Nardi, James B. *Once Upon a Tree: Life from Treetop to Root Tips.* Iowa City, IA: Iowa State University Press, 1993.

National Park Service, Katmai National Park and Preserve. http://www.nps.gov/katm/

Niering, William A. *Wetlands.* New York: Alfred A. Knopf, 1924.

"Ocean." *Encyclopedia Britannica.* Chicago: Encyclopedia Britannica, Inc., 1993.

"Ocean and Sea." *Grolier Multimedia Encyclopedia.* Danbury, CT: Grolier, Inc., 1995.

The Ocean. Scientific American, Inc., San Francisco: W. H. Freeman and Company, 1969.

Oceans. The Illustrated Library of the Earth. Emmaus, PA: Rodale Press, Inc., 1993.

Ocean World of Jacques Cousteau. Guide to the Sea. New York: World Publishing, 1974.

"On Peatlands and Peat." International Peat Society, http://www.peatsociety.fi

Oram, Raymond F. *Biology: Living Systems.* Westerville, OH: Glencoe/McGraw-Hill, 1994.

Owen, Andy, and Miranda Ashwell. *Lakes.* Portsmouth, NH: Heinemann Library, 1998.

Ownby, Miriam. *Explore the Everglades.* Kissimmee, FL: Teakwood Press, 1988.

Parker, Steve. *Eyewitness Books: Pond and River.* New York: Alfred A. Knopf, 1988.

Page, Jake. *Planet Earth: Forest.* Alexandria, VA: Time-Life Books, 1983.

Pernetta, John. *Atlas of the Oceans.* New York: Rand McNally, 1994.

Pipes, Rose. *Grasslands.* Orlando, FL: Raintree Steck-Vaughn Publishers, 1998.

Pipes, Rose. *Tundra and Cold Deserts.* Orlando, FL: Raintree Steck-Vaughn Publishers, 1999.

Plants of the Desert. Broomall, PA: Chelsea House Publishers, 1997.

Pringle, Laurence. *Fire in the Forest: A Cycle of Growth and Renewal.* New York: Atheneum Books, 1995.

Pringle, Laurence. *Rivers and Lakes.* Planet Earth. Alexandria, VA: Time-Life Books, 1985.

"Pripet Marshes" Britannica Online. http://www.eb.com:180/cgi-bin/g?DocF=micro/481/91.html

Pritchett, Robert. *River Life.* Secaucus, NJ: Chartwell Books, Inc., 1979.

Quinn, John R. *Wildlife Survival.* New York: Tab Books, 1994.

Radley, Gail, and Jean Sherlock. *Grasslands and Deserts.* Minneapolis, MN: The Lerner Publishing Group, 1998.

Renault, Mary. *The Nature of Alexander.* New York: Pantheon Books, 1975.

Rezendes, Paul, and Paulette Roy. *Wetlands: The Web of Life.* Vermont: Verve Editions, 1996.

Ricciuti, Edward. *Biomes of the World: Chaparral.* New York: Benchmark Books, 1996.

Ricciuti, Edward. *Biomes of the World: Grassland.* New York: Benchmark Books, 1996.

Ricciuti, Edward R. *Biomes of the World: Ocean.* New York: Benchmark Books, 1996.

"River and Stream." *Grolier Multimedia Encyclopedia.* Danbury, CT: Grolier, Inc., 1995.

"River." *Encyclopaedia Britannica.* Chicago: Encyclopaedia Britannica, Inc., 1993.

Rootes, David. *The Arctic.* Minneapolis: MN: Lerner, 1996.

Rosenblatt, Roger. "Call of the Sea." *Time* (October 5, 1998): 58-71.

Rotter, Charles. *The Prairie.* Mankato, MN: Creative Education, 1994.

Rotter, Charles. *Wetlands.* Mankato, MN: Creative Education, 1994.

Rowland-Entwistle, Theodore. *Jungles and Rainforests*. Our World. Morristown, NJ: Silver Burdett Press, 1987.

Sage, Bryan. *The Arctic and Its Wildlife*. New York: Facts on File, 1986.

Sanderson, Ivan. *Book of Great Jungles*. New York: Julian Messner, 1965.

Savage, Stephen. *Animals of the Desert*. Orlando, FL: Raintree Steck-Vaughn Publishers, 1996.

Savage, Stephen. *Animals of the Grasslands*. Orlando, FL: Raintree Steck-Vaughn Publishers, 1997.

"Saving the Salmon." *Time* (March 29, 1999): 60-61.

Sayre, April Pulley. *Desert*. Exploring Earth's Biomes. New York: Twenty-First Century Books, 1994.

Sayre, April Pulley. *Grassland*. New York: Twenty-First Century Books, 1994.

Sayre, April Pulley. *Lake and Pond*. New York: Twenty-First Century Books, 1996.

Sayre, April Pulley. *River and Stream*. New York: Twenty-First Century Books, 1995.

Sayre, April Pulley. *Seashore*. New York: Twenty-First Century Books, 1996.

Sayre, April Pulley. *Taiga*. New York: Twenty-First Century Books, 1994.

Sayre, April Pulley. *Temperate Deciduous Forest*. New York: Twenty-First Century Books, 1994.

Sayre, April Pulley. *Tundra*. New York: Twenty-First Century Books, 1994.

Sayre, April Pulley. *Wetland*. New York: Twenty-First Century Books, 1996.

Schoonmaker, Peter S. *The Living Forest*. New York: Enslow, 1990.

Schwartz, David M. *At the Pond*. Milwaukee, WI: Gareth Stevens, Inc., 1999.

Scott, Michael. *The Young Oxford Book of Ecology*. New York: Oxford University Press, 1995.

"Serengeti National Park" Britannica Online. http://www.eb.com:180/cgi-bin/g?DocF=micro/537/95.html

Sheldrake, Rupert. *A New Science of Life: The Hypothesis of Formative Causation*. Los Angeles: J. P. Tarcher, Inc., 1981.

Sheldrake, Rupert. *Seven Experiments That Could Change the World*. New York: Riverhead Books, 1995.

Shepherd, Donna. *Tundra*. Danbury, CT: Franklin Watts, 1997.

Silver, Donald M. *One Small Square: Arctic Tundra*. New York: W. H. Freeman and Company, 1994.

Silver, Donald M. *One Small Square: Pond*. New York: W. H. Freeman and Company, 1994.

Silver, Donald M. *One Small Square: Woods*. New York: McGraw-Hill, 1995.

Simon, Noel. *Nature in Danger*. New York: Oxford University Press, 1995.

Siy, Alexandra. *Arctic National Wildlife Refuge*. New York: Dillon Press, 1991.

Siy, Alexandra. *The Great Astrolabe Reef*. New York: Dillon Press, 1992.

Siy, Alexandra. *Native Grasslands*. New York: Dillon Press, 1991.

Slone, Lynn. *Ecozones: Arctic Tundra*. Viro Beach, FL: Rourke Enterprises, 1989.

"Soil" Microsoft Encarta 97 Encyclopedia. © 1993-1996 Microsoft Corporation.

Staub, Frank. *America's Prairies*. Minneapolis, MN: Carolrhoda Books, Inc., 1994.

Staub, Frank. *America's Wetlands*. Minneapolis, MN: Carolrhoda Books, Inc., 1995.

Staub, Frank. *Yellowstone's Cycle of Fire*. Minneapolis, MN: Carolrhoda Books, 1993.

Steele, Phillip. *Geography Detective: Tundra*. Minneapolis, MN: Carolrhoda Books, Inc., 1996.

Steele, Phillip. *Grasslands.* Minneapolis, MN: The Lerner Publishing Group, 1997.

Stidworthy, John. *Ponds and Streams.* Nature Club. Mahwah, NJ: Troll Associates, 1990.

Stille, Darlene R. *Grasslands.* Danbury, CT: Children's Press, 1999.

Storer, John. *The Web of Life.* New York: New American Library, 1953.

Sutton, Ann, and Myron Sutton. *Wildlife of the Forests.* New York: Harry N. Abrams, Inc., 1979.

Swink, Floyd, and Gerould Wilhelm. *Plants of the Chicago Region.* Lisle, IL: The Morton Arboretum, 1994.

Taylor, David. *Endangered Grassland Animals.* New York: Crabtree Publishing, 1992.

Taylor, David. *Endangered Wetland Animals.* New York: Crabtree Publishing, Co., 1992.

"Taymyr." Britannica Online. http://www.eb.com:180/cgi-Bin/g?DocF=micro/584/53.html

Thompson, Gerald, and Jennifer Coldrey. *The Pond.* Cambridge, MA: MIT Press, 1984.

Thompson, Wynne. *Journey to a Wetland.* Huntington, WV: Aegina Press, Inc., 1997.

"Thunderstorms and Lightning: The Underrated Killer." U.S. Department of Commerce. National Oceanic and Atmospheric Administration. National Weather Service, 1994.

"Tornado." Britannica Online. http://www.eb.com:180/cgi-bin/g?DocF=micro/599/23.html

"Tree." *Encyclopaedia Britannica.* Chicago: Encyclopaedia Britannica, Inc., 1993.

University of Wisconsin. "Stevens Point." http://www.uwsp.edu/acaddept/geog/faculty/ritter/geog101/biomes_toc.html

Wadsworth, Ginger. *Tundra Discoveries.* Walertown, MA: Charlesbridge Publishing, Inc., 1999.

Warburton, Lois. *Rainforests.* Overview Series. San Diego, CA: Lucent Books, Inc., 1990.

Waterlow, Julia. *Habitats: Deserts.* New York: Thomson Learning, 1996.

Waterflow, Julia. *Habitats: Grasslands.* Austin, TX: Raintree Steck-Vaughn Publishers, 1997.

Watson, Lyall. *Dark Nature.* New York: HarperCollins Publishers, 1995.

Wells, Susan. *The Illustrated World of Oceans.* New York: Simon and Schuster, 1991.

Wetlands: Meeting the President's Challenge. U.S. Dept. of the Interior, U.S. Fish and Wildlife Service, 1990.

"Wetlands." World Book Encyclopedia. Chicago: Scott Fetzer, 1996.

Whitfield, Philip, ed. *Atlas of Earth Mysteries.* Chicago: Rand McNally, 1990.

"Wind Cave National Park." http://www. nps.gov/wica/

Yates, Steve. *Adopting a Stream: A Northwest Handbook.* Seattle, WA: University of Washington Press, 1989.

Zich, Arthur. "Before the Flood: China's Three Gorges." *National Geographic.* (September, 1997): 2-33.

ORGANIZATIONS

African Wildlife Foundation
1400 16th Street NW, Suite 120
Washington, DC 20036
Phone: 202-939-3333
Fax: 202-939-3332
Internet: http://www.awf.org

American Cetacean Society
PO Box 1391
San Pedro, CA 90731
Phone: 310-548-6279
Fax: 310-548-6950
Internet: http://www.acsonline.org

American Littoral Society
Sandy Hook

BIBLIOGRAPHY

Highlands, NJ 07732
Phone: 732-291-0055

American Oceans Campaign
725 Arizona Avenue, Suite. 102
Santa Monica, CA 90401
Phone: 800-8-OCEAN-0
Internet: http://www.americanoceans.
com

American Rivers
1025 Vermont Avenue NW, Suite 720
Washington, DC 20005
Phone: 800-296-6900
Fax: 202-347-9240
Internet: http://www.amrivers.org

Astrolabe, Inc.
4812 V. Street, NW
Washington, DC 20007

Canadian Lakes Loon Survey
Long Point Bird Observatory
PO Box 160
Port Rowan, ON N0E 1M0

Center for Environmental Education
Center for Marine Conservation
1725 De Sales St. NW, Suite 500
Washington, DC 20036
Phone: 202-429-5609

Center for Marine Conservation
1725 De Sales Street, NW
Washington, DC 20036
Phone: 202-429-5609
Fax: 202-872-0619

Chihuahuan Desert Research Institute
PO Box 905
Fort Davis, TX 79734
Phone: 915-364-2499
Fax: 915-837-8192
Internet: http://www.cdri.org

Coast Alliance
215 Pennsylvania Ave., SE, 3rd floor
Washington, DC 20003
Phone: 202-546-9554
Fax: 202-546-9609
Internet: coast@igc.apc.org

Desert Protective Council, Inc.
PO Box 3635
San Diego, CA 92163
Phone: 619-298-6526
Fax: 619-670-7127

Environmental Defense Fund
257 Park Ave. South
New York, NY 10010
Phone: 800-684-3322
Fax: 212-505-2375
Internet: http://www.edf.org

Environmental Network
4618 Henry Street
Pittsburgh, PA 15213
Internet: http://www.envirolink.org

Environmental Protection Agency
401 M Street, SW
Washington, DC 20460
Phone: 202-260-2090
Internet: http://www.epa.gov

Forest Watch
The Wilderness Society
900 17th st. NW
Washington, DC 20006
Phone: 202-833-2300
Fax: 202-429-3958
Internet: http://www.wilderness.org

Freshwater Foundation
2500 Shadywood Rd.
Excelsior, MN 55331
Phone: 612-471-9773
Fax: 612-471-7685

Friends of the Earth
1025 Vermont Ave. NW, Ste. 300
Washington, DC 20003
Phone: 202-783-7400
Fax: 202-783-0444

Global ReLeaf
American Forests
PO Box 2000
Washington, DC 20005
Phone: 800-368-5748
Fax: 202-955-4588
Internet: http://www.amfor.org

Global Rivers Environmental Education
Network
721 E. Huron Street
Ann Arbor, MI 48104

Greenpeace USA
1436 U Street NW
Washington, DC 20009
Phone: 202-462-1177
Fax: 202-462-4507

Internet:
http://www.greenpeaceusa.org

International Joint Commission
1250 23rd Street NW, Suite 100
Washington, DC 20440
Phone: 202-736-9000
Fax: 202-736-9015
Internet: http://www.ijc.org

International Society of Tropical
Foresters
5400 Grosvenor Lane
Bethesda, MD 20814

Rainforest Action Movement
430 E. University
Ann Arbor, MI 48109

Isaak Walton League of America
SOS Program
1401 Wilson Blvd., Level B
Arlington, VA 22209

National Project Wet
Culbertson Hall
Montana State University
Bozeman, MT 59717

National Wetlands Conservation Project
The Nature Conservancy
1800 N Kent Street, Suite 800
Arlington, VA 22209
Phone: 703-841-5300
Fax: 703-841-1283
Internet: http://www.tnc.org

Nature Conservancy
1815 North Lynn Street
Arlington, VA 22209
Phone: 703-841-5300
Fax: 703-841-1283
Internet: http://www.tnc.org

North American Lake Management
Society
PO Box 5443
Madison, WI 53705
Phone: 608-233-2836
Fax: 608-233-3186
Internet: http://www.halms.org

Project Reefkeeper
1635 W Dixie Highway, Suite 1121
Miami, FL 33160

Rainforest Alliance
650 Bleecker Street

New York, NY 10012
Phone: 800-MY-EARTH
Fax: 212-677-2187
Internet: http://www.rainforest-
alliance.org

River Network
PO Box 8787
Portland, OR 97207
Phone: 800-423-6747
Fax: 503-241-9256
Internet: http://www.rivernetwork.org/
rivernet

Sierra Club
85 2nd Street, 2nd fl.
San Francisco, CA 94105
Phone: 415-977-5500
Fax: 415-977-5799
Internet: http://www.sierraclub.org

U.S. Fish and Wildlife Service Publication Unit
1717 H. Street, NW, Room 148
Washington, DC 20240

World Meteorological Organization
PO Box 2300
41 Avenue Guiseppe-Motta
1211 Geneva 2, Switzerland
Phone: 41 22 7308411
Fax: 41 22 7342326
Internet: http://www.wmo.ch

World Wildlife Fund
1250 24th Street NW
Washington, DC 20037
Phone: 202-293-4800
Fax: 202-293-9211
Internet: http://www.wwf.org

WEBSITES

Note: Website addresses are frequently
subject to change.

Arctic Studies Center: http://www.nmnh.
si.edu/arctic

Discover Magazine: http://www.
discover.com

Journey North Project: http://www.
jnorth@learner.org

Long Term Ecological Research Network: http://www.lternet.edu

BIBLIOGRAPHY

Monterey Bay Aquarium:
http://www.mbayaq.org

Multimedia Animals Encyclopedia:
http://www.mobot.org/MBGnet/vb/
ency.htm

National Center for Atmospheric
Research: http://www.dir.ucar.edu

National Geographic Magazine:
http://www.nationalgeographic.com

National Oceanic and Atmospheric
Administration: http://www.noaa.gov

National Park Service:
http://www.nps.gov

National Science Foundation:
http://www.nsf.gov/

Nature Conservancy:
http://www.tnc.org

Scientific American Magazine:
http://www.scientificamerican.com

Ouje-Bougoumou Cree Nation:
http://www.ouje.ca

Thurston High School, "Biomes"
http://www.ths.sps.lane.edu/biomes/
index1.html

Time Magazine: http://time.com/heroes

Tornado Project: http://www.
tornadoproject.com

University of California at Berkeley:
http://www.ucmp.berkeley.edu/
glossary/gloss5/biome/index.html

World Meteorological Organizaion:
http://www.wmo.ch

INDEX

Italic type indicates volume number; **boldface** type indicates entries and their pages; (ill.) indicates illustrations.

Saltwater crocodile *3:* 415

Saltwater lakes and ponds *2:* 220

Saltwater marshes *3:* 408, 484

Saltwater swamp *3:* 408, *3:* 482

Saltwater wetlands *3:* 408

Sami *3:* 459, 459 (ill.)

Samoans *3:* 424, 425 (ill.)

Sand *1:* 130; *2:* 225

Sand dunes *1:* 129, 131 (ill.); *3:* 407, 408 (ill.)

Sandgrouse *1:* 146

Sand hoppers *3:* 414

Saps *1:* 111

Sargasso Sea *2:* 261, 282

Savannas *2:* 178

Scorpions *1:* 140

Sea *2:* 261; *3:* 399

Sea anemone *1:* 66; *2:* 278, 278 (ill.)

Seabirds *1:* 64; *2:* 285; *3:* 417

Seafloor *1:* 55; *2:* 260, 270, 272

Seagrasses *1:* 60

Sea level *1:* 52; *2:* 227, 261; *3:* 398

Seals *1:* 66

Seamounts *2:* 271

Seashore *3:* **397–433**

   California *3:* 427

   Maine *3:* 428

   Netherlands *3:* 429

   North Carolina *3:* 429

   Norway *3:* 430

   Puerto Rico *3:* 427

   Texas *3:* 428

Seashore zones *3:* 402, 403 (ill.)

Sea snakes *1:* 63; *2:* 279

Seasons *1:* 90

Sea turtles *1:* 63, 66; *2:* 279; *3:* 415–16, 417 (ill.), 419

Seaweed *1:* 69; *3:* 410, 420

Secondary succession *1:* 7, 88; *2:* 309

Sedges *2:* 187; *3:* 491

Sediment *1:* 54, 72–73; *2:* 225–26, 245, 273, 292; *3:* 381–82, 423, 427

Seiche *2:* 223

Seif dunes *1:* 130

Semiarid regions *1:* 126

Seminole Indians *3:* 505, 509

Serengeti National Park *2:* 204

Serengeti plain *2:* 179, 180 (ill.)

Sewage *1:* 71; *3:* 378, 423

Shackleton, Ernest *3:* 398

Sharks *1:* 64; *2:* 283, 287

Shelf reefs *1:* 56

Sherpas *3:* 455, 460

Shoals *1:* 58; *2:* 225; *3:* 360

Shorebirds *2:* 239; *3:* 372, 417, 498

Siberian salamander *3:* 450

Sidewinders *1:* 144

Silver dragon *3:* 401

Silver Springs *2:* 219

Silviculture *1:* 4

Sinkholes *2:* 218

Skinks *1:* 24

Slash and burn agriculture *1:* 112; *2:* 330, 333 (ill.)

Sloth *2:* 326

Slough grass *2:* 187

Smokers *2:* 271, 277–78

Snakes *1:* 102, 144, 147 (ill.); *2:* 323

Soda lakes *2:* 220, 231

Soda ponds *2:* 220

Sodium carbonate *2:* 220

Sodium chloride *2:* 220

Softwoods *1:* 31, 109; *2:* 329

Soil *1:* 12, 91, 108, 130; *2:* 184, 229, 311, 328; *3:* 363, 438, 443, 487, 489

Solar energy *1:* 69; *2:* 288

Sonar *2:* 291

Soredia *1:* 14; *3:* 445

Sorghum *3:* 502

South China Sea *3:* 401

Southeast Asia *1:* 75

Southern Hemisphere Forest *1:* 11, 42

Speke, John Hanning *2:* 250

Sperm whale *1:* 68

Sphagnum moss *3:* 484–85, 485 (ill.)

Spiders *1:* 140

Spits *2:* 225; *3:* 358, 407

Splash zone *3:* 409

Spores *3:* 410, 445

Springtail *3:* 448

Spring tides *3:* 401

Squids *2:* 279